T0172536

An Introduction to
Abstract
Analysis

To Anita

An Introduction to
Abstract
Analysis

W.A. Light

Senior Lecturer
Mathematics Department
Lancaster University

CHAPMAN & HALL/CRC

Boca Raton   London   New York   Washington, D.C.

Library of Congress Cataloging-in-Publication Data

Catalog record is available from the Library of Congress.

First edition 1990
First CRC Press reprint 1999
Originally published by Chapman & Hall

© 1990 by W.A. Light

No claim to original U.S. Government works
International Standard Book Number 0-412-31080-5 (HB) 1 2 3 4 5 6 7 8 9 0
International Standard Book Number 0-412-31090-2 (PB) 1 2 3 4 5 6 7 8 9 0
Printed on acid-free paper

Contents

List of notations

Introduction

There are two objectives in this introduction. Firstly, I want to give a brief overview of the history of analysis as it relates to the material in this book. Secondly, I want to explain how the book is constructed, and how it is intended to be read.

The modern theory of analysis has its roots in the work of Liebnitz and Newton as they studied and developed the theory of differential and integral calculus in the 17th century. Central to this theory is the concept of a function. However, the modern definition of a function which every undergraduate mathematician encounters early in her undergraduate career is in fact the result of a long process of refinement. In the first half of the 17th century, functions were defined only by algebraic operations. Thus $f(s) = s^2$ was a typical function. The importance of the range, the domain and the rule defining the function was at this stage not evident. Also, a consequence of the algebraic nature of the definition was that knowledge of the function on a small interval allowed one to deduce the nature of the function for arbitrary values of the argument. To see what this means, suppose we know that a function f mapping real numbers to real numbers is defined by an algebraic rule. If we also know that the rule for real numbers in $[0,1]$ is given by $f(s) = s^2$, then, because the rule is known to be given by a single algebraic formula on the whole of $\mathbb{R}$, this rule must continue to hold for all real numbers. It must have universal truth. Even the idea that a function could be represented by different rules on different intervals of the real line was not thought of at this stage. The next significant step was the appearance of logarithmic and trigonometric functions. However, these functions still had a definition which had universal truth. Through the study of infinite series, the concept of a function began slowly to widen towards the end of the 18th and the beginning of the 19th centuries, and the idea of a function having several radically different forms of behaviour in different regions of the real line steadily gained acceptance.

In the 19th century, men like Cauchy, Weierstrass, Dirichlet and Riemann began to place the subject of functions on a firm abstract foundation. Their concept of a function and associated ideas such as continuity and differentiablity were essentially those which we still use today. However, their work did not meet with uniform approval amongst mathematicians of the day. The problem was that the class of objects which were now refered to as functions was widened tremendously by these founding fathers of analysis. This process of abstraction was eventually found to have granted citizenship to some fairly bizarre functions – ones which not only had no simple rule defining them, but could not even be drawn graphically.

Once the concept of a function was firmly established, the ideas of continuity and differentiability could be placed on a firm footing. This led to another controversy. There had been a strong feeling, brought about no doubt by considering the sorts of functions which can be graphed, that every continuous function was differentiable everywhere, or at least with the exception of a small number of 'special' points. As the abstract theory became better understood, both Bolzano and Weierstrass realised independently that this feeling was badly incorrect. (The reader who is interested in knowing just how badly wrong this feeling is, need only read as far as chapter 5 in this book.) Not everybody was thrilled with this state of affairs, however. Their work caused Hermite to express himself as follows: "I turn away with fear and horror from this lamentable plague of functions which do not have derivatives".

By the time the 20th century arrived, the foundations of the theory of functions of a real variable were well-understood, and analysis was ready for another change of pace and direction. Perhaps a major influence on this development was the pioneering work of G. Cantor at the end of the 19th century. What Cantor may be credited with achieving is bringing the abstract concept of a set to the forefront of mathematics. The theory of functions could then take another quantum leap to the modern position where the domain and range were abstract sets. The pace now quickened, largely due to the French school of mathematicians. The work of Borel and Lebesgue on techniques for measuring sets opened the way for an abstract theory of integration which proved to be far superior to the original ideas of Riemann. Baire and Borel were involved in a theory of classification for functions, and many other able researchers contributed to making the beginning of the 20th century a period of rapid development. Running parallel, and often inextricably intertwined with the real theory was the theory of functions of a complex variable.

With the publication of a book called *Théorie des opérations linéaires* [1] by a Polish mathematician, S. Banach, a new abstract framework in which many of the previous ideas could be discussed was created. It soon became apparent that this framework (that of normed linear spaces – see chapter 1) offered powerful tools to the rest of analysis. In addition, there was considerable interest in the subject for its own sake, and this interest has given rise to important modern areas of analysis such as operator theory and the geometric theory of Banach spaces. At the same time, Banach's work gave mathematicians what amounts to a new language for analysis. Such was the power of this new language, that today it is well-understood by mathematicians whose involvement in abstract mathematics is quite small. For example, this language is prevalent in the field of numerical analysis – where techniques for using computers as aids to problem solving are studied.

This presents an interesting pedagogical problem. Anyone wishing to work in the analytical side of mathematics must be familiar with this modern language. But when is the correct time to introduce such material? A student who has only received an informal training in the methods of calculus (so that no formal proofs have been encountered) would find the combination of the formality of analysis and the strangeness of the abstract framework just too daunting. However, once a student has done the equivalent of one semester of univariate formal calculus, encountering the concepts of sequences, limit, continuity, differentiability and integration, and most important, has had plenty of practice with rigorous proofs using quantifiers, then the next step can involve some gentle introduction to normed linear spaces. Of course, a small amount of knowledge about linear structures is also necessary, but much can be done with nothing more than an understanding of the concept of a linear space together with the ideas of dimensionality and basis.

A major problem with this scheme of things is that most books in this field are simply too ambitious for the student who is still coming to terms with the formal notions of analysis. The majority of authors set out with the intention of covering a substantial portion of the theory at a late undergraduate or beginning postgraduate level. This often requires advanced technical machinery, such as topological spaces, and dictates a pace which is far too rapid for someone who is still coming to terms with the way in which analysts think, and the process by which results are established. This book attempts to provide a gentle introduction to the theory of normed linear spaces, while at the same time exposing the way in which the common arguments of analysis work. How is this achieved? Firstly, the pace of exposition at the start of the book is slow. (At least, in my opinion, it is slow. Experience has shown me that I do not always succeed in striking the right pace in the eyes of students!) I have tried to take considerable care in the initial chapters to point out the basic strategies of proof, and the common pitfalls. As the chapters progress, I presume that the reader is gradually becoming familiar with the subject, through her reading and through her working of the exercises. Consequently, the pace quickens, and the reader is left more and more to her own devices. Secondly, the choice of topics has been severely limited, so that this book is not of daunting length, neither does it contain many of the great landmarks in the subject. I considered most of these just too difficult to tackle in an introductory work. In addition, I have looked very much to the field of function theory for most of the examples. In doing so, a severe blow has been dealt to a large part of the subject, which deals with linear spaces where the objects are sequences or Lebesgue measurable functions. I feel that functions which have domain and range in the real numbers are objects which most readers will feel reasonably at home with, and so this is the best place to go for practical outlets of the theory. It has the added advantage that the applications in this book continue the themes that the student will have encountered in previous courses – continuity, ideas from calculus, polynomials, and so on.

One of the consequences of my approach is that the reader really should go on after reading this book to read other, more advanced books. I very much hope that she will do so. One of my main objects in writing this book has been to try to convey my enthusiasm for the subject to the reader, and I will regard myself as having failed if this is the last book on analysis which the reader studies.

Now a word about the structure of the book. Each chapter contains exercises liberally sprinkled throughout. My intention is that the reader should do most, if not all of these. None are really hard, and many round out the treatment of the theory. Some are even essential for certain future arguments. All theorems, lemmas, corollaries are numbered consecutively within the chapter they occur. The proofs are set off in the text by the head **Proof**, and are terminated by a small black box.

I have benefited greatly from the teaching I have received and from the many discussions I have had with colleagues about the correct way to teach analysis. It is a pleasure to acknowledge that assistance here. I am greatly indebted to my friend and colleague, Professor E. W. Cheney whose friendship and collaboration over many years has served to deepen my understanding of the subject, and to stimulate my own research interests in the area. Dr. G. J. O. Jameson has also had considerable influence on my teaching of analysis during the 18 years we have been colleagues, and this book owes much to my perusal of his analysis lecture notes. Professor W. Deeb was somehow persuaded to read the manuscript when it was almost in final form. I benefited greatly from his gentle but persistent criticisms.

Finally, I owe a deep debt of gratitude to my wife, Anita, who became a 'computer widow', while I struggled with the writing and typesetting of this book. This book is dedicated to her.

Will Light
Lancaster, 1990.

1
Basic ideas

We shall begin with some ideas which are fundamental to the rest of this book. Basic to all of mathematics is the idea of a *function* and the associated notation. We shall use the notation $f : A \to B$ to denote the fact that A and B are two sets and f is a function (mapping) between them. It is usually understood that A is the domain of f so that $f(a)$ is the unique element of B corresponding to a via the mapping f. Things are a little different with regard to the set B, however. We normally consider B to contain the range of the function but strict containment is permissible, so that all of B need not necessarily be 'used up' by the function. Notice that each function has three pieces of information. Firstly, the domain A; secondly, the set B, which contains the range of the function; and, thirdly the rule – how to get from A to B using the mapping f. For example, the rule $f(x) = 1/x$ with $A = (0, \infty)$ and $B = \mathbb{R}$ is quite satisfactory, whereas the same rule with $A = [0, \infty)$ and $B = \mathbb{R}$ is not. (The destination of the point $x = 0$ under the rule is not defined.) Similarly, the rule $f(x) = \sin x$ with $A = B = \mathbb{R}$ is satisfactory, as is the same rule with $A = \mathbb{R}$ and $B = [-1, 1]$. In the second case the set B (or the 'target space') is exactly the range of the function.

Having made such a fuss over the care needed in talking about functions made up of the three ingredients rule, domain and target space, we often abuse the notation by refering to 'the function f'. Here the domain and target space are omitted, and this is frequently done when both are clearly understood from the context. This is usually the case when $A = B = \mathbb{R}$ or when A and B have been mentioned previously and it would be belabouring the point to continually write

$f : A \to B$. All careful mathematicians are aware of the fact that omitting the domain and/or the target space is fraught with danger and therefore try to abuse the notation carefully!

Our second fundamental ingredient is that of a linear space. The set X is a linear space if there is some method by which any two elements of X can be 'added together' to give a third member of X, and each element of X can be 'multiplied' by a real number to give another element of X. For example, the Cartesian plane $\mathbb{R}^2$ is a linear space. The points in $\mathbb{R}^2$ consist of coordinate pairs (s, t) and 'addition' means

$$(s, t) + (s_1, t_1) = (s + s_1, t + t_1).$$

Similarly, 'multiplication' by the real number (scalar) α means

$$\alpha(s, t) = (\alpha s, \alpha t).$$

Of course, the processes called addition and multiplication are not the simple concepts used in $\mathbb{R}$, although they are very similar. For example,

$$(s, t) + (s_1, t_1) = (s + s_1, t + t_1) = (s_1, t_1) + (s, t),$$

so that the new concept of addition is commutative, as is addition in $\mathbb{R}$. Furthermore, the origin $(0, 0)$ has the property that

$$(0, 0) + (s, t) = (s, t) = (s, t) + (0, 0).$$

Thus the point $(0, 0)$ in the linear space $\mathbb{R}^2$ plays an analogous role to that of 0 in $\mathbb{R}$. For this reason, it is often referred to as the *zero element* in the linear space $\mathbb{R}^2$. Given any point (s, t) in $\mathbb{R}^2$, the point $(-s, -t)$ has the property that

$$(s, t) + (-s, -t) = (0, 0) = (-s, -t) + (s, t).$$

The point $(-s, -t)$ is usually called the additive inverse of (s, t).

Consider the set X consisting of all mappings from $\mathbb{R}$ into $\mathbb{R}$. This is also a linear space. The elements in X consist of functions, and addition of two functions is effected by defining their sum to be the 'pointwise sum', so that if f and g are two points in X (i.e. two functions from $\mathbb{R}$ to $\mathbb{R}$) then the function $h = f + g$ is defined by

$$h(s) = f(s) + g(s), \quad s \in \mathbb{R}.$$

Similarly, the function y which is the result of multiplying f by the scalar α is defined by

$$y(s) = \alpha f(s), \quad s \in \mathbb{R}.$$

Faced with a concrete situation it is usually easy to say what is meant by addition and by multiplication by scalars. However, the description of these two properties in an abstract setting is rather harder. Notice that 'addition' associates with every pair of elements x, y in X a third element z which we call $x + y$. Association in mathematics usually involves mappings, and 'addition' is a mapping from $X \times X$ (ordered pairs of elements in X) into X with certain properties that make the mapping 'look like' the usual process of addition of real numbers. In a similar way scalar multiplication is a mapping from $\mathbb{R} \times X$ (ordered pairs of elements, the first element lying in $\mathbb{R}$ and the second in X) into X. It will have certain properties that make it resemble multiplication in $\mathbb{R}$. Now we are ready to say formally what constitutes a linear space.

Definition 1.1 *A linear space is a set X together with two mappings $\phi : X \times X \to X$ and $\psi : \mathbb{R} \times X \to X$ such that*

1. $\phi(x, y) = \phi(y, x)$ *for* $x, y \in X$

2. $\phi(x, \phi(y, z)) = \phi(\phi(x, y), z)$ *for* $x, y, z \in X$

3. *there exists a unique element θ in X such that $\phi(x, \theta) = \phi(\theta, x) = x$ for x in X*

4. *to each element x in X there corresponds a unique element y such that $\phi(x, y) = \theta$*

5. $\psi(\alpha, \phi(x, y)) = \phi(\psi(\alpha, x), \psi(\alpha, y))$ *for* $\alpha \in \mathbb{R}$, $x, y \in X$

6. $\phi(\psi(\alpha, x), \psi(\beta, x)) = \psi(\alpha + \beta, x)$ *for* $\alpha, \beta \in \mathbb{R}$, $x \in X$

7. $\psi(\alpha, \psi(\beta, x)) = \psi(\alpha\beta, x)$ *for* $\alpha, \beta \in \mathbb{R}$, $x \in X$

8. $\psi(1, x) = x$ *for* $x \in X$.

After the simplicity of the examples the formality looks quite daunting. However, we almost never use the mappings ϕ and ψ. Instead we always think of the mapping ϕ as 'addition' and write it as $\phi(x, y) =$

$x + y$. This is fine, as long as we remember that the sign '+' is now being used for a whole host of different meanings, depending on the context in which it is being employed. The above conditions now read

1. $x + y = y + x$ for $x, y \in X$

2. $x + (y + z) = (x + y) + z$ for $x, y, z \in X$

3. there exists a unique element θ in X such that $x + \theta = \theta + x = x$, for all $x \in X$

4. to each x in X there corresponds a unique element y (which we write as $-x$) such that $x + y = x + (-x) = \theta$

5. $\alpha(x + y) = \alpha x + \alpha y$ for $\alpha \in \mathbb{R}$, $x, y \in X$

6. $\alpha x + \beta x = (\alpha + \beta)x$ for $\alpha, \beta \in \mathbb{R}, x, y \in X$

7. $\alpha(\beta x) = (\alpha \beta)x$ for $\alpha, \beta \in \mathbb{R}, x \in X$

8. $1x = x$, for $x \in X$.

In these conditions we can see quite clearly the different meanings of 'addition'. For example, in condition 6 the addition on the left of the equality sign represents addition in the linear space, whereas the addition on the right represents the usual addition in $\mathbb{R}$. From now on we will always refer to $\phi(x, y)$ as $x + y$ and to $\psi(\alpha, x)$ as αx. Of course, in defining a linear space we need to decide three things; what choice we will make for the set of objects or points, how we will form the sum of two points and how we will take products of a point in the set and a real number. In fact the full impact of the axioms for a linear space as given in **1.1** is rarely realised in analysis. We usually work with linear spaces in which there is a simple 'natural' definition of addition and scalar multiplication, and so the problem of determining whether a given set is a linear space is rarely of great importance. Such problems belong more properly to the field of algebra. The words linear space convey the sense of the linear structure of these objects, but the historical development sometimes leads to the alternative phrase vector space.

   Analysis is bound up with the idea of 'closeness' and our next concept introduces a measure of distance in a linear space. Let us return to our example $X = \mathbb{R}^2$. We are familiar with the idea that

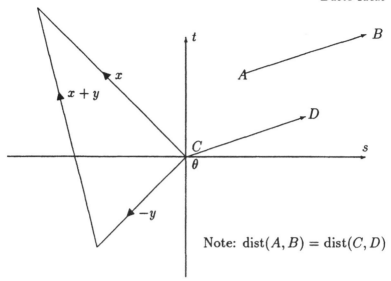

Note: dist(A, B) = dist(C, D)

Figure 1.1: Distances in $\mathbb{R}^2$.

the distance between two points $x_1 = (-2, 1)$ and $x_2 = (1, 2)$ is $\sqrt{([1 - (-2)]^2 + [2 - 1]^2)} = \sqrt{10}$. Notice in Figure 1.1 that $\sqrt{10}$ is also the distance from the point $(3, 1)$ to the origin $\theta = (0, 0)$, and that $(3, 1) = (1, 2) - (-2, 1)$. This is not an accident! It will always happen that one of the properties we will require of our notion of distance is that the distance from a point x_1 to a point x_2 in $\mathbb{R}^2$ will be the same as the distance from $x_1 - x_2$ or $x_2 - x_1$ to the origin. Note that such a requirement rests partly on the fact that $\mathbb{R}^2$ is a linear space (otherwise we could never talk about $x_1 - x_2 = x_1 + (-x_2)$). The distance between points in $\mathbb{R}^2$ will therefore be completely determined once we know the distance of each individual point from the origin. There are two further properties of the distance which we will require. Both properties describe how the notion of distance ties in with the linear structure. For example, it would be nice if the distance of $10x$ from the origin was 10 times the distance of x from the origin. Thus if $\alpha \in \mathbb{R}$ then we shall demand that the distance of αx from the origin θ is $|\alpha|$ times the distance of x from θ. The modulus has appeared here because we always want distance to be a non-negative real number. This describes how the concept of distance interacts with that of multiplication by a scalar in the linear space structure. How should

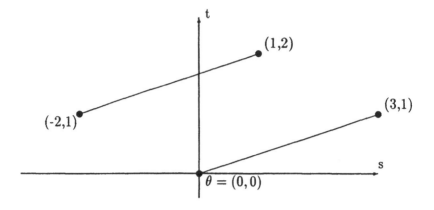

Figure 1.2: Distance remains constant under a simple linear shift.

this concept interact with that of addition of elements of the linear space? Roughly speaking, we want our measurement of distance to ensure that the direct distance between two points $x, y \in \mathbb{R}^2$ never exceeds the sum of the distances between x and any intermediate point z and z and y. Because of the linear space structure, this is the same as requiring that the distance of $x + y$ from θ is never bigger than the sum of the individual distances of x and y from θ. This last fact is illustrated in Figure 1.1. Notice that the concept of distance has associated each point in $\mathbb{R}^2$ with a real number – its distance from the origin. Once again this association is expressed mathematically by the idea of a mapping.

Definition 1.2 *Let X be a linear space. Then a norm is a mapping $\rho : X \to \mathbb{R}$ such that*

   1. $\rho(x) \geq 0$ for all x in X and $\rho(x) = 0$ if and only if $x = \theta$

   2. $\rho(\alpha x) = |\alpha|\rho(x)$ for $\alpha \in \mathbb{R}, x \in X$

   3. $\rho(x + y) \leq \rho(x) + \rho(y)$ for $x, y \in X$.

As with the addition and multiplication maps it is unusual to encounter the mapping ρ. We nearly always write $\rho(x)$ as $\|x\|$ so that the above conditions read

   1. $\|x\| \geq 0$ for all x in X and $\|x\| = 0$ if and only if $x = \theta$

2. $\|\alpha x\| = |\alpha|\|x\|$ for $\alpha \in \mathbb{R}, x \in X$

3. $\|x + y\| \leq \|x\| + \|y\|$ for $x, y \in X$.

From the preceding remarks for $\mathbb{R}^2$ we see that a suitable choice of norm for that linear space is

$$\|x\| = \|(s, t)\| = \sqrt{(s^2 + t^2)}.$$

The whole of the rest of this book is bound up with the ideas of linear spaces and norms. A linear space X on which a norm is defined will be called a *normed linear space* and written $(X, \|\cdot\|)$. Once again, we often abuse this notation when the norm is clear from the context by refering to 'the normed linear space' X. It will become clear as we go on through the subject that certain linear spaces have natural norms. An example of this is the norm given above for $\mathbb{R}^2$. It corresponds to our natural notion of distance in $\mathbf{R}^2$. There are other norms which can be defined on this linear space, but they give us notions of distance in $\mathbb{R}^2$ which do not correspond to our natural intuition.

There are several helpful pointers which will enable you to understand the structure of a specific normed linear space. Firstly, it is a good idea to make sure you understand thoroughly the notions of addition and scalar multiplication in the linear space. Then it is a good idea to try to visualise some simple sets which are defined by reference to the given norm. For example, the sets

$$A = \{x \in X : \|x\| \leq 1\}, \quad U(y, \epsilon) = \{x \in X : \|x - y\| \leq \epsilon\}$$

for fixed y in X and fixed ϵ in $\mathbf{R}$. The next set of exercises investigates some of the common normed linear spaces that we will encounter, and the experience of this investigation is not to be missed.

Exercises

1. Show that the following all define norms on $\mathbf{R}^2$:

    (a) $\|x\|_1 = \|(s, t)\|_1 = |s| + |t|$

    (b) $\|x\|_\infty = \|(s, t)\|_\infty = \max\{|s|, |t|\}$

    (c) $\|x\|_2 = \|(s, t)\|_2 = \sqrt{(s^2 + t^2)}$.

    Sketch the set A and some examples of the sets $U(y, \epsilon)$.

2. Let $C[-1,1]$ be the linear space of continuous functions from $[-1,1]$ to $\mathbb{R}$. Show that the following defines a norm on this linear space;

$$\|x\| = \max\{|x(s)| : s \in [-1,1]\}.$$

Again, sketch the set A and some examples of the sets $U(y, \epsilon)$.

3. For the three definitions of norm on $\mathbb{R}^2$ given in question 1, compute the distance from the point $(1,1)$ to $(2,3)$.

4. Give an example of elements a and b in $\mathbb{R}^2$ such that $\|a\|_1 < \|b\|_1$, while $\|a\|_2 > \|b\|_2$. (The norms $\|\cdot\|_1$ and $\|\cdot\|_2$ are defined in question 1.)

5. Let X be the linear space consisting of all bounded real-valued functions on $[0, 2\pi]$. Take $\|x\| = \sup\{|x(s)| : s$ lies in $[0, 2\pi]\}$. Take x and y in X defined by $x(s) = \cos s$ and $y(s) = \sin s$. Calculate $\|x - y\|$.

6. Let X, x and y be as in the previous exercise. Show that

$$\|\lambda f + \mu g\| = (\lambda^2 + \mu^2)^{\frac{1}{2}}, \quad \text{for all } \lambda, \mu \text{ in } \mathbb{R}.$$

[Hint: there exits an α such that $\cos \alpha = \lambda/(\lambda^2 + \mu^2)^{\frac{1}{2}}$.]

7. Let X be the set of $n \times n$ matrices with real entries. Take the natural definitions of addition and scalar multiplication. For a matrix $A = (a_{ij})$ set

$$\|A\| = \max\{\sum_{i=1}^{n} |a_{ij}| : 1 \le j \le n\}.$$

Show that X together with this definition of norm forms a normed linear space. It is too hard to sketch the usual sets in this case!

8. Let $X = C[a, b]$. Define a norm on X by

$$\|x\| = \int_a^b |x(s)| \, ds.$$

Show that this really does define a norm on $C[a, b]$, and sketch the set A and some examples of the sets $U(y, \epsilon)$. If $a = 0$, $b = 1$, x and y are given by $x(s) = s$ and $y(s) = -s^2 + 1$, compute $\|x - y\|$.

Notice that there are two distinct types of linear space here. There is the *finite-dimensional* linear space $\mathbb{R}^2$. For such a space we can obtain a *basis*, in the sense that each element of $\mathbb{R}^2$ can be written as a linear combination of the two basis elements $(0,1)$ and $(1,0)$. These basis elements possess another property in addition to the one just mentioned: they are *linearly independent.* That is, if α and β are in $\mathbb{R}$ and

$$\alpha(0,1) + \beta(1,0) = \theta = (0,0),$$

this implies that $\alpha = \beta = 0$. This choice of basis elements is not unique. For example the pairs $(2,0)$, $(0,3)$ or $(1,2)$, $(1,3)$ are equally satisfactory. An important invariant here is the number of basis elements, which is always 2, and for this reason $\mathbb{R}^2$ is said to have *dimension* 2.

In general, a linear space X is said to be n-dimensional if the largest number of linearly dependent elements in the space is n. In this case, every basis for X contains n elements, and each element of X can be written uniquely as a linear combiation of the basis elements. Thus if $\{x_1, \ldots, x_n\}$ is a basis for X, and $x \in X$, then there exist unique real numbers $\lambda_1, \ldots, \lambda_n$ such that

$$x = \sum_{i=1}^{n} \lambda_i x_i.$$

The space $C[-1,1]$ is, however, quite different in that we cannot find such a basis for this space, because it is just too big. To understand this, observe that many linear spaces contain within them subsets which are linear spaces in their own right. Such subsets are called *subspaces*. A simple example in $C[-1,1]$ is the subset consisting of all functions which are polynomials of degree at most one. Such functions are of the form $x(s) = a + bs$ where $a, b \in \mathbb{R}$. It is clear that if we add two such functions, or if we multiply such a function by a scalar, then we generate another function of the same type (a polynomial of degree at most one). A basis for this subspace consists of the two functions y and z given by $y(s) = 1$ and $z(s) = s$, $s \in [-1,1]$, and so this subspace has dimension 2. It is now clear that $C[-1,1]$ contains many similar subspaces. For example, the subspace consisting of polynomials of degree at most n is an n-dimensional subspace of $C[-1,1]$. This space is often called *infinite-dimensional* because it has no basis in the sense that there is no finite subset of $C[-1,1]$ whose linear combinations generate the whole space. This follows from the fact that such a finite subset would have to generate all polynomials of every possible degree, which is manifestly impossible.

Exercises

1. Let X be the set of all real sequences $\{x_n\}_1^\infty$ such that $\sum_{i=1}^\infty |x_i|$ is convergent. Show that

$$\|x\| = \sum_{i=1}^\infty |x_i|$$

   defines a norm on X. Note here that each point x in X is a sequence $\{x_i\}_{i=1}^\infty$.

2. Let Y be the set of all real, bounded sequences. Show that

$$\|y\| = \sup\{|y_i| : 1 \le i < \infty\}$$

   defines a norm on Y.

3. Let Z be the set of all real sequences with limit zero. Show that Z together with the norm from the previous question is a normed linear space.

4. Show that the spaces X, Y and Z as defined in the previous questions are infinite-dimensional.

5. Let X be the linear space of all $n \times n$ matrices. Discover the dimension of X by exhibiting a basis for X.

2

Some simple results

So far we have defined two very simple structures – that of a linear space and its associated norm. The beauty and power of abstract analysis is that these two concepts are completely adequate for us to talk about many of the familiar topics in real analysis. For example, we can make sense of the notions of convergence of sequences and continuity of functions. We shall define precisely what is meant by these notions in our abstract setting and then explore the implications of these definitions. In addition, we shall introduce some special concepts which help to illuminate the ones just mentioned. Our first definition is one such concept.

Definition 2.1 *The ball centre x and radius r in the normed linear space X is the set $\{y : \|x - y\| \leq r\}$. This is sometimes called the closed ball, the open ball being the set $\{y : \|x - y\| < r\}$.*

Figure 2.1 shows typical open and closed balls in $\mathbb{R}^2$ with the norm $\|(s, t)\|_1 = |s| + |t|$. It is convention that when the boundary is included in the set (the closed ball) we draw a solid boundary while if the boundary is not included (the open ball) we draw a broken boundary.

Definition 2.2 *A sequence $\{x_n\}$ in a normed linear space X converges to $x \in X$ if given $\epsilon > 0$ we can find N (depending on ϵ) such that $\|x - x_n\| < \epsilon$ for all $n > N$. The point x is called the limit of the sequence.*

We sometimes use the notation $x_n \to x$ or $\lim x_n = x$. It is sometimes helpful to paraphrase this definition as follows: whatever size of open

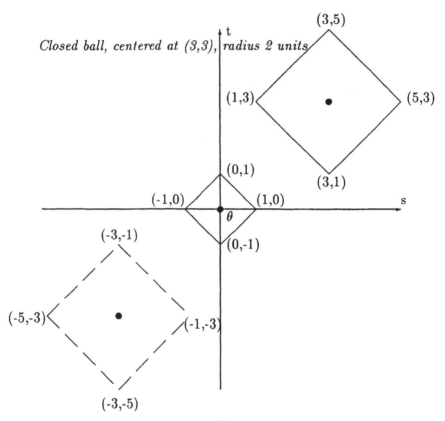

Figure 2.1: Balls in R^2 with the norm $\| \cdot \|_1$.

ball we pick with centre x there is some place in the sequence after which all subsequent members lie in that open ball. One of the early results for real sequences is that the limit of a sequence is unique. We give the corresponding result for normed linear spaces.

Lemma 2.3 *If a sequence $\{x_n\}$ of elements in a normed linear space X converges to a limit, then that limit is unique.*

Proof. Suppose that $x_n \to x$ and $x_n \to y$ where x and y are in X. Then using the triangle inequality we obtain

$$\|x - y\| = \|x - x_n + x_n - y\| \leq \|x - x_n\| + \|x_n - y\|.$$

There are now two ways of proceeding. The professional analyst might say that since the above inequality must hold for all n, it must hold in the limit and so $\|x - y\| = 0$. Then it follows from the first property of a norm that $x = y$, and so the limit is unique. However, to see what the phrase 'in the limit' really means we need to resort to **2.2**. Given any $\epsilon > 0$ we can find an N such that $\|x_n - x\| < \epsilon$ and $\|x_n - y\| < \epsilon$ for all $n > N$. Hence for all such values of n we have $\|x - y\| < 2\epsilon$ in the above inequality. Since this last inequality is true for all positive ϵ, we conclude that $\|x - y\| = 0$ and so $x - y = \theta$, that is $x = y$. ∎

Notice that fundamental to the above proof was the use of the triangle inequality to obtain the displayed inequality as a basic tool. The other ingredient is the explanation of how this tool is used to achieve the desired result. This is common to many of the proofs in analysis. A simple inequality, or string of inequalities, provides a basic tool which is then employed via a logical argument which consists mostly of text accompanied by mathematical symbolic abbreviations.

Lemma 2.4 *If a sequence $\{x_n\}$ in a normed linear space X converges to the point x in X and $\|x_n\| \leq M$ for all n, then $\|x\| \leq M$.*

Proof. We prove the required result by contradiction. Accordingly, assume $x_n \to x$, $\|x_n\| \leq M$ and $\|x\| > M$. We can assume that $\|x\| = M + \alpha$ where $\alpha > 0$. Since $x_n \to x$, there exist an N such that $\|x_n - x\| < \alpha/2$ for all $n > N$. Now,

$$\|x\| = \|x - x_n + x_n\| \leq \|x - x_n\| + \|x_n\| \leq \|x - x_n\| + M.$$

If we take $n > N$ then we obtain $\|x\| \leq M + \alpha/2$, which contradicts our assumption that $\|x\| = M + \alpha$. ∎

Again, notice the fundamental inequality surrounded by the supporting argument. Another feature of this argument is what could be described as the 'insertion-deletion principle'. This is the process of writing the element x in X as $x+y-y$, thus inserting and deleting the element y. We normally use this principle to make an insertion and deletion which combine with existing terms to give quantities about which we have information. In the above proof we had information about the terms $x - x_n$ and x. Accordingly, we inserted and deleted the term x_n.

Another way to discover the strategy needed for a particular proof is to consider a simple example. Often the choice $X = \mathbb{R}^2$ with the usual definition of norm, $\|(s,t)\| = \sqrt{s^2 + t^2}$, will suffice. Figure 2.2 indicates how the proof of **2.4** could be arrived at from a suitable picture, once the decision to use the technique of proof by contradiction has been adopted. Of course, pictures in themselves prove nothing. They simply serve as an aid to abstract thinking.

This same 'insertion-deletion' principle can be used to obtain a useful consequence of the triangle inequality.

Lemma 2.5 *Let x and y belong to the normed linear space X. Then*

$$\left| \|x\| - \|y\| \right| \leq \|x - y\| \leq \|x\| + \|y\|.$$

Proof. The right-hand inequality is the familiar triangle inequality for the norm and is restated here for convenience only. To establish that the modulus of a number is at most $\|x - y\|$ we must show that it lies between $-\|x - y\|$ and $\|x - y\|$, i.e.

$$-\|x - y\| \leq \|x\| - \|y\| \leq \|x - y\|.$$

The right-hand inequality follows from

$$\|x\| = \|x - y + y\| \leq \|x - y\| + \|y\|,$$

while the left-hand side follows from

$$\|y\| = \|y + x - x\| \leq \|y - x\| + \|x\|. \qquad \blacksquare$$

Note that the left-hand inequality in **2.5** is a *stronger* statement than $\|x\| - \|y\| \leq \|x - y\|$, although this last statement is true since

$$\|x\| - \|y\| \leq \left| \|x\| - \|y\| \right| \leq \|x - y\|.$$

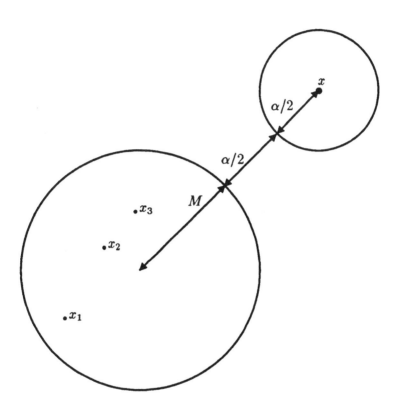

Figure 2.2: Illustration of the proof of 1.4 in $\mathbb{R}^2$

Basically, **2.5** says that if we take the smaller of $\|x\|$ and $\|y\|$ from the larger then this (non-negative) difference is bounded above by $\|x - y\|$. We do not always use the full strength of **2.5**, however. Sometimes it suffices to use the weaker condition $\|x\| - \|y\| \le \|x - y\|$.

Definition 2.6 *A set A in a normed linear space X is bounded if there exists a number $M > 0$ such that $\|a\| \le M$ for all a in A.*

Lemma 2.7 *If A and B are bounded sets in a normed linear space then so is $A \cup B$.*

Proof. (Sketch) Any element x in $A \cup B$ has $\|x\| \le M$ where M is the maximum of the individual bounds for A and B. ■

Suppose we have a mapping $f : X \to Y$ where $(X, \|\cdot\|_X)$ is one normed linear space while $(Y, \|\cdot\|_Y)$ is another. Here we have temporarily expanded our notation to indicate the space to which each norm is refering. One of the desirable events in the behaviour of a function is that small changes in the location of x in the domain space should produce a small change in the location of $f(x)$ in the range space. Such a function is said to be continuous.

Definition 2.8 *Let $(X, \|\cdot\|_X)$ and $(Y, \|\cdot\|_Y)$ be normed linear spaces. Then a function $f : X \to Y$ is continuous at x_0 in X if, given an $\epsilon > 0$ there is a $\delta > 0$ such that $\|x_0 - x\|_X < \delta$ implies $\|f(x_0) - f(x)\|_Y < \epsilon$. If the function f is continuous at all points of X, then f is said to be continuous.*

The above definition is not easy to handle for a variety of reasons. Firstly, it is what we call a 'two quantifier' statement, the quantifiers being ϵ and δ. Essential to an understanding of continuity is an awareness of the roles that ϵ and δ play in **2.8**. This will be particularly important when we come to describe a function which is not continuous at some point in X. Secondly, it does not quite say what we would expect from the loose idea of continuity which immediately preceded the definition. The formal definition focusses attention on the range space rather than the domain. Thirdly, the technicalities of the definition make it difficult to use in practice. We get around this by constructing a handful of continuous functions and then deriving rules for composing them in various ways to get other continuous functions. A nice class of functions from $\mathbb{R}$ to $\mathbb{R}$ which are continuous is the class of differentiable functions. We are not yet in a position to talk about differentiability of functions whose domain is a normed linear space, but the K-Lipschitz functions can be thought of as an analogous class to the differentiable functions. Let K be a positive real number. A function $f : X \to Y$ is said to be K-Lipschitz if

$$\|f(x_1) - f(x_2)\|_Y \le K\|x_1 - x_2\|_X \quad \text{for all } x_1 \text{ and } x_2 \text{ in } X.$$

It is easy to see that a K-Lipschitz function is continuous. To prove this, fix x_0 in X. We must show that to each $\epsilon > 0$ there corresponds a $\delta > 0$ such that $\|f(x) - f(x_0)\|_Y < \epsilon$ whenever $\|x - x_0\|_X < \delta$. Whenever we want to show that a statement is true for all real numbers $\epsilon > 0$, we always proceed in the same way. We take a real number $\epsilon > 0$. Then as long as we do not place any further restrictions on it, the number ϵ will be playing the role of a general, or generic, positive, real number. Thus any conclusions we draw will hold for *all* positive, real numbers. We must be careful, of course, not to assume later on in the proof that ϵ is somehow special. An unsubtle example of this would be if we suddenly saw that our argument would only work if we assumed $\epsilon > 1$. The moment we write, 'now assume $\epsilon > 1$' we can no longer hope to prove our assertion for *all* real numbers $\epsilon > 0$. Of course, such mistakes are usually more subtle than this. For example, we may at some point divide by the quantity $\epsilon - 1$. This would have the consequence that none of the subsequent arguments are valid for $\epsilon = 1$. So, take $\epsilon > 0$ and let x_0 belong to X. How small must we choose δ so that $\|f(x) - f(x_0)\|_Y < \epsilon$ whenever $\|x - x_0\|_X < \delta$? The answer can be read off from the Lipschitz inequality. If we take $\delta \le \epsilon/K$, then for all x satisfying $\|x - x_0\|_X < \delta$ we have

$$\|f(x) - f(x_0)\|_Y \le K\|x - x_0\|_X < K\delta \le K.\frac{\epsilon}{K} = \epsilon.$$

It will be convenient to omit the subscript from the norms and allow the reader to deduce which norm is intended from the context. Otherwise, large parts of the ensuing text would become submerged in a mass of clumsy notation.

Lemma 2.9 *Let X, Y and Z be normed linear spaces. If $f : X \to Y$ and $g : Y \to Z$ are continuous mappings then $g \circ f$ is a continuous mapping from X into Z.*

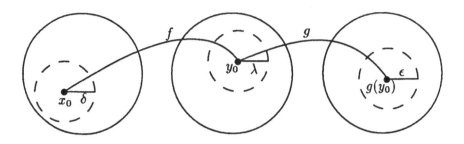

Figure 2.3: Composition of continuous functions.

Proof. Fix x_0 in X. We must show that for each $\epsilon > 0$ there is a $\delta > 0$ such that $\|g(f(x)) - g(f(x_0))\| < \epsilon$ whenever $\|x - x_0\| < \delta$. Write $y_0 = f(x_0)$. Since g is continuous at y_0 there exists $\lambda > 0$ such that if $\|y - y_0\| < \lambda$ then $\|g(y) - g(y_0)\| < \epsilon$. Since f is continuous at x_0, there exists $\delta > 0$ such that if $\|x - x_0\| < \delta$ then $\|f(x) - f(x_0)\| < \lambda$ and so $\|g(f(x)) - g(f(x_0))\| < \epsilon$. ∎

When proofs are written out formally, we can often follow the logical argument from step to step, but we may not have a clear idea of the motivation behind the structuring of the argument. Previously, we drew a picture of a simple example. It is also possible to draw diagrams which illustrate proofs without necessarily tying us down to a specific example. Figure 2.3 illustrates the motivation behind the previous proof.

We now come to an alternative characterisation of continuity. The result is quite profound when compared with anything that we have done so far. Thus we can expect to encounter significant difficulties in the proof.

Lemma 2.10 *Let f be a mapping of a normed linear space X into another normed linear space Y. Then f is continuous at x_0 if and only if whenever $\{x_n\}$ is a sequence in X with limit x_0, we have $f(x_n) \rightarrow f(x_0)$.*

Proof. We do the easy direction first. Suppose f is continuous at x_0 and let ϵ be a fixed positive number. There is a $\delta > 0$ such that if $\|x - x_0\| < \delta$ then $\|f(x) - f(x_0)\| < \epsilon$. Furthermore, if $\{x_n\}$ is a sequence with limit x_0 then there is an $N > 0$ such that $\|x_n - x_0\| < \delta$

for all $n \geq N$. Thus $\|f(x_n) - f(x_0)\| < \epsilon$ for all $n \geq N$, and this is precisely the property that is required for the sequence $\{f(x_n)\}$ to have limit $f(x_0)$. The other half of the proof is effected by a 'contrapositive' argument. Thus in trying to prove the assertion (statement p) $\Rightarrow$ (statement q), we shall in fact establish the equivalent assertion that (negation of q)$\Rightarrow$ (negation of p). The main difficulty lying before us is the negation of the appropriate statements. We shall begin by supposing that f is not continuous at x_0 . For f to be continuous we need to achieve a certain task for every $\epsilon > 0$. If f is not continuous then we must fail at this task for at least one $\epsilon > 0$. Call this value ϵ_0. What does it mean to fail at the task for this value ϵ_0? If we look back to the definition of continuity (**2.8**), then we will fail in our task if we cannot find a $\delta > 0$ such that $\|x - x_0\| < \delta$ implies $\|f(x) - f(x_0)\| < \epsilon_0$. Thus whatever value of $\delta > 0$ we choose, there must exist a point x_δ such that $\|x_0 - x_\delta\| < \delta$ but $\|f(x_\delta) - f(x_0)\| \geq \epsilon_0$. Now take a sequence $\{\delta_n\}$ such that $\delta_n \to 0$. Then (altering our notation slightly by writing x_n for x_{δ_n}) to each δ_n there corresponds a point x_n with $\|x_n - x_0\| < \delta_n$ but $\|f(x_n) - f(x_0)\| \geq \epsilon_0$. This means that $x_n \to x_0$ but that $f(x_n) \not\to f(x_0)$. In summary, the fact that f is not continuous at x implies that there is a convergent sequence $\{x_n\}$ with limit x_0 such that $\{f(x_n)\}$ does not converge to $f(x_0)$. This is precisely the contrapositive assertion. ∎

There are two important points that arise out of the above result and its proof. Firstly, the technique of contrapositive argument is very important in mathematics and you should become very familiar with it. Usually, such an argument involves the negating of rather complex statements, and we need to exercise special care whenever this arises. Secondly, it is often easier to use **2.10** than the original definition of continuity. This raises the question as to why we did not define continuity via **2.10** immediately. If $x_n \to x_0$ implies $f(x_n) \to f(x_0)$ then we say f is sequentially continuous at x. However, the concepts of continuity and sequential continuity are not always equivalent. Rather this is a special feature of normed linear spaces and there are other settings in which continuity may be defined in accordance with **2.8** but this definition is not equivalent to sequential continuity.

Notice that in the definition of a function $f : X \to Y$ being continuous it is conceivable that for a given $\epsilon > 0$ the δ needed may get

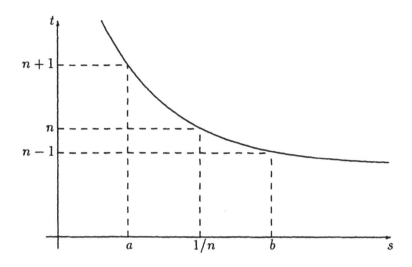

Figure 2.4: The function $f(x) = 1/x$ near $x = 0$.

very small as we shift our attention from point to point in X. This underlines the fact that the parameter δ is dependent on the point x in the definition. For example, consider the case $X = Y = \mathbb{R}$ with the norm $\|x\| = |x|$. Take $f : X \to Y$ defined by $f(x) = 1/x$ for $x \neq 0$ and $f(0) = 0$. Then this function is continuous everywhere except $x = 0$. Now fix ϵ at the value 1 and consider the type of δ needed to satisfy the definition of continuity at the points $y = 1/n$, $n \in \mathbb{N}$. We want to find δ such that if

$$|x - 1/n| < \delta \quad \text{then} \quad |f(x) - f(1/n)| < 1.$$

Using the definition of f we see that we want δ to satisfy

$$|x - 1/n| < \delta \quad \text{implies} \quad |(1/x) - n| < 1.$$

In figure 2.4 we need to calculate the values of a and b. Then the largest possible value for δ will be the smallest of $1/n - a$ and $b - 1/n$. We have

$$\frac{1}{n} - a = \frac{1}{n} - \frac{1}{n+1} = \frac{1}{n^2 + n},$$

while

$$b - \frac{1}{n} = \frac{1}{n-1} - \frac{1}{n} = \frac{1}{n^2 - n}.$$

Thus the largest possible value we may take for δ is $1/(n^2 + n)$. As we take y nearer to zero (i.e., n larger), we see that we require a smaller and smaller value for δ in order to retain the inequality $|f(x) - f(1/n)| < 1$. More important is the fact that there is no single value of δ which will work for *all* values of y. This turns out to be a major disadvantage in much of the theory, and the desirability of using functions for which a single value of δ holds for all points, either in the whole space or in some subset of the space, is the motivation for the following definition.

Definition 2.11 *Let f be a mapping between the normed linear spaces X and Y. We say that f is uniformly continuous on a set A contained in X if, given $\epsilon > 0$, there is a $\delta > 0$ such that whenever x_1 and x_2 are in A and $\|x_1 - x_2\| < \delta$, then $\|f(x_1) - f(x_2)\| < \epsilon$.*

The function f is therefore uniformly continuous on a set A in X if it is firstly continuous at every point of A, and if, secondly, for each value of $\epsilon > 0$ there is a common or uniform value of δ which suffices in the definition of continuity. Thus uniform continuity is a stronger requirement than continuity. As with continuity, uniform continuity is not an easy property to handle. However, one consequence of some rather powerful theorems which come later in this book will be that all continuous functions are uniformly continuous on a wide class of subsets of a normed linear space. For example, we shall show that every continuous function from $\mathbb{R}$ to $\mathbb{R}$ is uniformly continuous on each closed interval in $\mathbb{R}$.

The previous discussion can now be used to see that the function $f : \mathbb{R} \to \mathbb{R}$ defined by $f(x) = 1/x$ for $x \neq 0$ and $f(0) = 0$ is not uniformly continuous on the set $A = (0, 1)$. The mapping $f(x) = x^2$ between the same two linear spaces is uniformly continuous on A, since we may consider for any value of $\delta > 0$

$$|(y \pm \delta)^2 - y^2| = |\pm 2y\delta + \delta^2| \leq 2\delta|y| + \delta^2 \leq 2\delta^2 + \delta.$$

Thus taking $\delta > 0$ such that $2\delta + \delta^2 < \epsilon$ will ensure that

$$|f(x) - f(y)| < \epsilon \quad \text{whenever} \quad |x - y| < \delta.$$

The class of K-Lipschitz functions from one normed linear space X to another, Y, is now seen to possess the further desirable property that each K-Lipschitz function is uniformly continuous on any set $A \subset X$, and of course on X itself.

Definition 2.12 *The distance of a point x in a normed linear space X from a set A in X is defined as*

$$\text{dist}(x, A) = \inf\{\|x - a\| : \ a \text{ is in } A\}.$$

Note that this definition extends our existing notion that the quantity $\|x - a\|$ measures the distance between the points x and a in the normed linear space. The present definition says that the distance from a point x in X to a set A is the infimum of the individual distances of x from points in A. Observe also that there need not be a point a^* in A for which $\|x - a^*\| = \text{dist}(x, A)$, that is, the infimum in **2.12** need not be attained. When such a point a^* does exist it is called a *closest point* or *best approximation* to x from A. For example if $X = \mathbb{R}$ with the usual definition of norm (i.e. the norm of a point is equal to its modulus) and $A = (0, 1)$, then $\text{dist}(2, A) = 1$, but there is no point a^* in $(0, 1)$ such that $\|x - a^*\| = 1$.

Lemma 2.13 *Let f be the function from the normed linear space X to $\mathbb{R}$ defined by $f(x) = \text{dist}(x, A)$ where A is some fixed set in X. Then f is a 1-Lipschitz function.*

Proof. Take x_1, x_2 in X and a in A. Then

$$\text{dist}(x_1, A) \le \|x_1 - a\| \le \|x_1 - x_2\| + \|x_2 - a\|.$$

Hence

$$\text{dist}(x_1, A) - \|x_1 - x_2\| \le \|x_2 - a\|.$$

Since this is true for all a in A we have

$$\text{dist}(x_1, A) - \|x_1 - x_2\| \le \text{dist}(x_2, A),$$

and so

$$\text{dist}(x_1, A) - \text{dist}(x_2, A) \le \|x_1 - x_2\|.$$

In a similar fashion we obtain

$$\text{dist}(x_2, A) - \text{dist}(x_1, A) \le \|x_1 - x_2\|,$$

and so

$$|\text{dist}(x_1, A) - \text{dist}(x_2, A)| \le \|x_1 - x_2\|.$$

This shows that our function f is 1-Lipschitz, as required. ∎

Exercises

1. Show that if f and g are bounded, K-Lipschitz functions from the normed linear space X to $\mathbb{R}$ then $f.g$ is Lipschitz. Give a counterexample when f and g are unbounded.

2. Show that if f is a K-Lipschitz function on the normed linear space X, then f is uniformly continuous on X.

3. Show that if X is a normed linear space and $x, y \in X$ with $\|x\| \leq 1$, $\|y\| \leq 1$, then

$$\|\alpha x + (1 - \alpha)y\| \leq 1, \quad 0 \leq \alpha \leq 1.$$

4. Let x and y be elements of a normed linear space with $\|x\| = 1$. Show that either

   (a) $\|x + \lambda y\| \geq 1$ for all $\lambda > 0$, or
   (b) $\|x - \lambda y\| \geq 1$ for all $\lambda > 0$.

   Hint: Suppose that both are false and use the identity

$$\|\alpha a + (1 - \alpha)b\| \leq |\alpha|\|a\| + |1 - \alpha|\|b\|.$$

5. If x and y are in a normed linear space X and $\|x+y\| = \|x\|+\|y\|$ prove that $\|x + \lambda y\| = \|x\| + \lambda\|y\|$, for all $\lambda > 0$. (Hint: there are two cases, $\lambda > 1$ and $\lambda \leq 1$.)

6. Let x, y belong to the normed linear space X, and $\|x+y\| \leq M$, $\|x - y\| \leq M$. Show that $\|x + \alpha y\| \leq M$ for all $-1 \leq \alpha \leq 1$.

7. Suppose in a normed linear space X, $x_n \to x$ and $y_n \to y$. Show that $\|x_n\| \leq \|y_n\|$ for all $n \in \mathbb{N}$ implies $\|x\| \leq \|y\|$. What can be concluded if $\|x_n\| < \|y_n\|$ for all $n \in \mathbb{N}$? What happens if $\|x_n\| = \|y_n\|$ for all $n \in \mathbb{N}$?

8. Let X be a normed linear space, and define the mapping u by $u(x) = x/\|x\|$, for $x \in X \setminus \{\theta\}$. Show that

$$\|u(x) - u(y)\| \leq \frac{2\|x - y\|}{\|x\|}.$$

   Deduce that u is continuous on $X \setminus \{\theta\}$.

9. Prove that the elements of a convergent sequence in a normed linear space always form a bounded set.

10. Let $\mathbb{R}$ have the usual norm. Calculate
    (i) $\text{dist}(1, \{t : t^2 > 2\})$
    (ii) $\text{dist}(0, \{t : t = 1/n, n \in \mathbb{N}\})$.

11. Let $X = \mathbb{R}^2$ with the norm defined by

$$\|(s, t)\| = |s| + |t|.$$

Let $A = \{(s, t) : s = t\}$. What is $\text{dist}(y, A)$, when $y = (0, 1)$? Describe the set of best approximations to y from A.

12. Let $X = C[-1, 1]$ with norm $\|x\| = \sup\{|x(t)| : t \in [-1, 1]\}$, for $x \in X$. Let A be the set of constant functions in $C[-1, 1]$. Let $x \in X$ be defined by $x(t) = t^2$. Compute $\text{dist}(x, A)$. Does x have a best approximation in A?

13. Let X be a normed linear space, and let f be a continuous function from X to $\mathbb{R}$. Show that the function $|f|$ defined by $|f|(x) = |f(x)|$ for $x \in X$ is continuous.

14. Prove that the following define continuous functions from D to $\mathbb{R}$:
    (i) $D = \{x \in \mathbb{R} : x \geq 0\}$ and $f(x) = \sqrt{x}$
    (ii) $D = \mathbb{R}$ and $f(s) = a_n s^n + a_{n-1} s^{n-1} + \cdots + a_0$, where $a_0, a_1, \ldots a_n$ are fixed real numbers.

15. Define $f : \mathbb{R} \to \mathbb{R}$ by

$$f(s) = \begin{cases} s & s \text{ irrational} \\ 1 - s & s \text{ rational.} \end{cases}$$

Show that f is continuous at $s = 1/2$ and discontinuous everywhere else.

16. Show that $f : \mathbb{R} \to \mathbb{R}$ defined by $f(s) = s^2$ is not uniformly continuous on $\mathbb{R}$.

17. Let X, Y, and Z be normed linear spaces. If $f : X \to Y$, is uniformly continuous on $S \subset X$ and $g : Y \to Z$ is uniformly continuous on $f(S) \subset Y$, show that $g \circ f$ is uniformly continuous on S.

18. Let $f : \mathbb{R} \rightarrow \mathbb{R}$ be continuous on the interval $[a, b]$. Define $g : [a, b] \rightarrow \mathbb{R}$ by $g(s) = \max\{f(t) : t \in [a, s]\}$ if $s \neq a$ and $g(a) = f(a)$. Show that g is continuous on $[a, b]$.

We shall conclude this section by examining the continuity of functions on the linear space $\mathbb{R}^n$. Recall that $\mathbb{R}^n$ is the set of n-tuples of the form

$$(x_1, x_2, x_3, \ldots, x_n)$$

where each entry x_i, $1 \leq i \leq n$, is a real number. The usual definitions of addition and multiplication by scalars are adopted, so that addition of two n-tuples is effected by adding the individual components and multiplication by a scalar simply means that each member of the n-tuple is multiplied by that scalar. Hence,

$$(x_1, x_2, \ldots, x_n) + (y_1, y_2, \ldots, y_n) = (x_1 + y_1, x_2 + y_2, \ldots, x_n + y_n)$$

and

$$\alpha(x_1, x_2, \ldots, x_n) = (\alpha x_1, \alpha x_2, \ldots, \alpha x_n).$$

There are three principal definitions which we adopt for the norm in $\mathbb{R}^n$. If we assume that $x = (x_1, x_2, \ldots, x_n)$ then these norms are defined as follows:

$$\|x\|_2 = \{\sum_{i=1}^{n} x_i^2\}^{\frac{1}{2}}, \quad \|x\|_1 = \sum_{i=1}^{n} |x_i|, \quad \|x\|_\infty = \max\{|x_i| : 1 \leq i \leq n\}.$$

It is perhaps a little surprising that the verification that the first of these definitions really has the properties required of a norm is quite difficult. The other two are straightforward, but one of the exercises guides you through the proof of the first case. There are sensible historical reasons for the subscripting of these norms, which come from a much larger family of norms on $\mathbb{R}^n$. Throughout this section the symbol $\| \cdot \|$ is used to denote any one of the three norms above and unless mention is made of a particular norm any statement about convergence or continuity should be understood to refer to each of the three norms. The reason for this ambiguity is contained in the next lemma.

Lemma 2.14 *For any x in $\mathbb{R}^n$ we have*
(i) $\|x\|_\infty \leq \|x\|_1 \leq n\|x\|_\infty$.
(ii) $\|x\|_\infty \leq \|x\|_2 \leq n\|x\|_\infty$.

Proof. These inequalities are easy to prove and we give only the proof of (i). Firstly,

$$\|x\|_1 = \sum_{i=1}^{n} |x_i| \geq \max_i |x_i| = \|x\|_\infty.$$

Secondly,

$$\|x\|_1 = \sum_{i=1}^{n} |x_i| \leq \sum_{i=1}^{n} \max_j |x_j| = n\|x\|_\infty. \qquad \blacksquare$$

We often summarise **2.14** by saying that the three norms are all equivalent. What this result is really saying is that whenever one of the three norms is small, so are the other two. This has the implication that if $\{x_n\}$ is a sequence in $\mathbb{R}^n$ with limit x with respect to one of the norms it will converge and have the same limit with respect to the other two norms. Similarly, if $f : \mathbb{R}^n \to X$ is continuous when $\mathbb{R}^n$ has one of the three norms, then it is continuous when $\mathbb{R}^n$ has either of the other two norms. Our chief interest in the next few results is in the connection between the behaviour of elements of $\mathbb{R}^n$ and the behaviour of the individual coordinates. To help us notationally, we sometimes refer to the ith coordinate projection on $\mathbb{R}^n$. This is the mapping $P_i : \mathbb{R}^n \to \mathbb{R}$ which maps the element $x = (x_1, x_2, \ldots, x_n)$ to the element x_i , the ith coordinate of x.

Lemma 2.15 *The ith coordinate projection is a continuous mapping from $\mathbb{R}^n$ to $\mathbb{R}$.*

Proof. It will be sufficient by the equivalence of the norms (**2.14**) to show the continuity when we use the norm $\|.\|_\infty$. In fact the mapping P_i is 1-Lipschitz as the following inequality shows:

$$|P_i x - P_i y| = |x_i - y_i| \leq \max_{1 \leq j \leq n} |x_j - y_j| = \|x - y\|_\infty. \qquad \blacksquare$$

Before we continue we need to overcome the notational problems involved with $\mathbb{R}^n$. The points in $\mathbb{R}^n$ will be labelled by single letters such as x. When we want to speak about the coordinates of x we will label these

$$x(1), x(2), \ldots, x(n),$$

so that

$$x = (x(1), x(2), \ldots, x(n)).$$

A sequence $\{x_k\}$ in $\mathbb{R}^n$ can then have $x_k(i)$ as the ith coordinate of the kth point in the sequence, so that

$$x_k = (x_k(1), x_k(2), ..., x_k(n)).$$

In a similar fashion any mapping f from a normed linear space X to $\mathbb{R}^n$ can have its action described by

$$f(x) = (f_1(x), f_2(x), \ldots, f_n(x))$$

where each f_i is uniquely defined by f and is a mapping from X into $\mathbb{R}$. An interesting question is whether properties of f such as continuity are transferred to the 'coordinate' functions f_i. In a well-behaved world they surely will be!

Lemma 2.16 *Let $\{x_k\}$ be a sequence of elements in $\mathbb{R}^n$. Then the following are equivalent:*

1. *$x_k \to x$ as $k \to \infty$.*

2. *For each i, $1 \le i \le n$, $x_k(i) \to x(i)$ as $k \to \infty$.*

Proof. Recall that there are really three separate statements here involving the three different norms on $\mathbb{R}^n$, but the equivalence of norms in **2.14** allows us to deal only with the case $\|\cdot\|_\infty$. Suppose initially that $\|x_k - x\|_\infty \to 0$ as $k \to \infty$. Then given $\epsilon > 0$ there is a K such that $\|x_k - x\|_\infty < \epsilon$ for all $k > K$. This means that

$$\max\{|x_k(i) - x(i)| : 1 \le i \le n\} < \epsilon \quad \text{for all } k > K.$$

Thus $|x_k(i) - x(i)| < \epsilon$ for each i, $1 \le i \le n$, and this is precisely what is required to conclude that $x_k(i) \to x(i)$ for each i, $1 \le i \le n$. Conversely, if $x_k(i) \to x(i)$ for $1 \le i \le n$, then given $\epsilon > 0$ there will exist numbers $K_1, K_2, \ldots, K_n$ such that $|x_k(i) - x(i)| < \epsilon$ whenever $k > K_i$. Now take $K = \max_i K_i$. Then for $1 \le i \le n$, $|x_k(i) - x(i)| < \epsilon$ whenever $k > K$. This will imply that

$$\max\{|x_k(i) - x(i)| : 1 \le i \le n\} < \epsilon \text{ for all } k > K,$$

and so $\|x_k - x\|_\infty \to 0$ as $k \to \infty$. $\blacksquare$

Lemma 2.17 *Let X be a normed linear space. Let $f : \mathbb{R}^2 \to X$ be continuous at the point (s_0, t_0). Define $g : \mathbb{R} \to X$ by $g(s) = f(s, t_0)$. Then g is continuous at s_0.*

Proof. This argument is simple if we use the sequential notion of continuity. Accordingly, let $\{s_n\}$ be a convergent sequence of real numbers with limit s_0. Then from **2.15** we have that $(s_n, t_0) \to (s_0, t_0)$ and so, since f is continuous at (s_0, t_0), we obtain $f(s_n, t_0) \to f(s_0, t_0)$. This may be rephrased to give the required result, namely $g(s_n) \to g(s_0)$. ∎

Lemma 2.18 *Let f be a mapping of a normed linear space X into $\mathbb{R}^n$. Then f is continuous at x in X if and only if each of the corresponding coordinate functions f_i, $1 \le i \le n$, is continuous.*

Proof. Suppose each f_i is continuous at x. Then given $\epsilon > 0$ there exists $\delta > 0$ such that $|f_i(x) - f_i(y)| < \epsilon$ whenever $\|x - y\| < \delta_i$ and $1 \le i \le n$. Now take $\delta = \min \delta_i$. Then, for $1 \le i \le n$, we have

$$|f_i(x) - f_i(y)| < \epsilon \text{ whenever } \|x - y\| < \delta.$$

Again it is sufficient to consider the norm $\|.\|_\infty$ on $\mathbb{R}^n$. The above inequality implies that

$$\|f(x) - f(y)\|_\infty = \max\{|f_i(x) - f_i(y)| : 1 \le i \le n\} < \epsilon,$$

whenever $\|x - y\| < \delta$. Hence f is continuous. For the other direction we simply observe that the mapping f_i can be written as the composition $P_i \circ f$. Then P_i is continuous by **2.15**, and f is continuous by hypothesis, so that f_i is continuous by **2.9**. ∎

Exercises

1. The first two exercises show how to verify that $\| \cdot \|_2$ defines a norm on $\mathbb{R}^n$. Define a mapping from $\mathbb{R}^n \times \mathbb{R}^n \to \mathbb{R}$ by

$$g(x, y) = \sum_{i=1}^{n} x_i y_i,$$

where $x = (x_1, x_2, \ldots, x_n)$ and $y = (y_1, y_2, \ldots, y_n)$. Show that
(i) $g(x, x) = \|x\|_2^2$ for all $x \in \mathbb{R}^n$
(ii) $g(x, y) = g(y, x)$ for all $x, y \in \mathbb{R}^n$
(iii) $g(x, y+z) = g(x, y) + g(x, z)$ and $g(x+y, z) = g(x, z) + g(y, z)$ for all $x, y, z \in \mathbb{R}^n$
(iv) $g(\lambda x, y) = \lambda g(x, y) = g(x, \lambda y)$ for all $x, y \in \mathbb{R}^n$ and $\lambda \in \mathbb{R}$.

2. With the same notation as the previous question show that

$$g(x,y) \leq \|x\|_2 \|y\|_2$$

by considering $g(z, z)$ where $z = \|y\|x - \|x\|y$. Use this property to verify the triangle inequality for $\| \cdot \|_2$.

3. Show that the following definitions for f define a continuous function from $\mathbb{R}^2$ to $\mathbb{R}^2$:
(i) $f(s,t) = (2s + t, s - 2t)$
(ii) $f(s,t) = (s^2 + t^2, 2st)$

4. Let $f, g : \mathbb{R}^n \to \mathbb{R}$ be continuous at a point $a \in \mathbb{R}^n$. Define $h, k : \mathbb{R}^n \to \mathbb{R}$ by

$$h(x) = \sup\{f(x), g(x)\} \quad \text{and} \quad k(x) = \inf\{f(x), g(x)\}, \quad x \in \mathbb{R}^n.$$

Show that h and k are continuous at a.

5. Let $F : \mathbb{R}^2 \to \mathbb{R}$ be defined by

$$F(s,t) = \begin{cases} s^2 + t^2 & s, t \in \mathbb{Q} \\ 0 & \text{otherwise.} \end{cases}$$

Determine the points where F is discontinuous.

6. Suppose $f, g : \mathbb{R}^n \to \mathbb{R}$ are continuous at $x \in \mathbb{R}^n$, and $g(x) \neq 0$. Show that f/g is continuous at x.

7. A function $f : \mathbb{R}^n \to \mathbb{R}$ is said to be additive if

$$f(x + y) = f(x) + f(y), \quad \text{for all } x, y \text{ in } \mathbb{R}^n.$$

Show that an additive function which is continuous at $x = \theta$ is continuous everywhere in $\mathbb{R}^n$.

3

Open and closed sets

In univariate real analysis the open and closed intervals given by
$(a, b) = \{s \in \mathbb{R} : a < s < b\}$ and $[a, b] = \{s \in \mathbb{R} : a \leq s \leq b\}$
play an important role in many theorems. A simple example of such
a theorem is that every continuous function mapping a closed interval $[a, b]$ into $\mathbb{R}$ is bounded and attains its bounds (i.e. it achieves
its supremum and infimum). The result is no longer true if we consider continuous functions defined on open intervals. The function
$f : (0, 1) \to \mathbb{R}$ defined by $f(s) = s^2$ is certainly continuous and
bounded. It is bounded above by 1 and below by 0. However, it does
not attain the lower bound zero. That is, there is no point s in $(0, 1)$
such that $f(s) = 0$. The function $f : (0, 1) \to \mathbb{R}$ defined by $f(s) = 1/s$
is again continuous. However, f is certainly not bounded. From such
elementary considerations we might expect to develop notions of open
and closed sets in a normed linear space, and to have these concepts
play an important role in the developing theory. This is indeed the
case, although we shall discover later that the natural generalisation of
the above result does not rest entirely on the notion of closure which
we shall introduce.

What differentiates between open and closed intervals in $\mathbb{R}$? Well,
there are a number of ways of expressing the differences. For example,
an open interval does not contain its endpoints, whereas a closed interval does. When stated this way, such a description is too dependent
on the univariate situation – we would find it rather difficult to decide
what consituted an 'endpoint' for a set in $\mathbb{R}^2$, say. If we want to generalise in this way, it would be more natural to regard the endpoints
of an interval as forming the boundary of the set. Then we could go

ahead and formalise a notion of boundary in a normed linear space. However, it is more elegant to consider other ways of describing open intervals in $\mathbb{R}$. Each point in an open interval (a, b) is 'interior' to that interval in the sense that we can move a small amount to either side of the point while remaining within (a, b). Of course, our 'freedom of movement' is restricted by how close our point is to either of the points a and b. (See figure 3.1.) The idea of being able to move a small amount to either side of a given point can be formulated more accurately by saying that there is an open interval centred on that point (i.e. if the given point is s_0, then a centred open interval would have the form $(s_0 - r, s_0 + r)$ for some $r > 0$: alternatively, we can write such an interval as $\{s : |s - s_0| < r\}$), which is contained in (a, b). For the closed interval $[a, b]$ the points a and b are clearly *not* interior to the set $[a, b]$.

We must decide now what is means for a point in a set A which is contained in a normed linear space X to be 'interior' to A. To do this we need the analogue of the open interval, and this is called the open ball. The *open ball* centred at a in X with radius $r > 0$ is denoted by $B_r(a)$ and defined by

$$B_r(a) = \{x \in X : \|x - a\| < r\}.$$

Definition 3.1 *Let A be a subset of a normed linear space X. The point a is an interior point of A if there is an $r > 0$ such that $B_r(a)$ is contained in A. The notation $\mathrm{int}\,A$ stands for the set of interior points of A. The set A is called open if each of its points is an interior point, i.e. if $A = \mathrm{int}\,A$.*

Note that if we want to establish that a certain set A is open using the above definition, we start by considering a general point a in A and proceed by showing how to construct the open ball $B_r(a)$ which is contained in A. This reduces to a question of deciding the appropriate radius. As an example, let us show that the set $B_r(a)$ is itself open. Figure 3.2 illustrates the next few lines of argument. Pick any point x_0 in $B_r(a)$. Then $\|x_0 - a\| = r_1 < r$. Set $r' = r - r_1$. Then $r' > 0$ and so $B'_r(x_0)$ is an open ball centred on x_0. We claim $B'_r(x_0)$ lies within our original ball $B_r(a)$, i.e. $B_{r'}(x_0) \subseteq B_r(a)$. To see this set inclusion holds, take $x \in B_{r'}(x_0)$. Then $\|x - a\| \leq \|x - x_0\| + \|x_0 - a\|$ by the

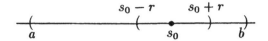

Figure 3.1: Freedom of movement in the set (a, b).

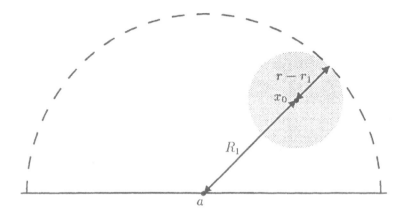

Figure 3.2: Illustration of the proof that $B_r(a)$ is open.

triangle inequality and so

$$\|x - a\| < r' + r_1 = r - r_1 + r_1 = r.$$

Since $\|x - a\| < r$, $x \in B_r(a)$ as required. Now we return to the idea of boundaries, albeit in a slightly disguised form.

Definition 3.2 *Let A be a subset of a normed linear space X and let $x \in X$. Then x is a point of closure of A if for every $\delta > 0$ there is an a in A with $\|x - a\| < \delta$. The set of all points of closure of A, written $\overline{A}$, is called the closure of A. If $A = \overline{A}$, we say A is closed.*

A consequence of this definition is that any point which is in A is also a point of closure of A. However, there may well be points outside A which are also points of closure of A. Such points can be visualised as 'sitting on the boundary' of A. Equivalently, a point of closure of A is a point x for which $\text{dist}(x, A) = \inf\{\|x - a\| : a \in A\} = 0$. As an example consider the set $A = [0, 1)$ as a subset of the normed linear space $\mathbb{R}$. Each point in $[0, 1)$ is a point of closure of $[0, 1)$ as remarked previously. However, the point 1 is also a point of closure of A. To see this, we have to take $\delta > 0$ and exhibit an a in A with $\|1 - a\| = |1 - a| < \delta$. The point $1 - \delta/2$ is a suitable choice for a. Thus the closure of $[0, 1)$ is $[0, 1]$, as we would expect. We might now go on to ask for the closure of $[0, 1]$. This is in fact $[0, 1]$ and so $[0, 1]$ is a closed set in $\mathbb{R}$. To see that $[0, 1]$ contains all its points of closure we must show that any point x outside $[0, 1]$ is not a point of closure of this set. Suppose $x > 1$. Then $x - 1 = r > 0$. All the points y in $\mathbb{R}$ such that $|x - y| < r/2$ have $y > 1$ and so no such point lies in $[0, 1]$. Thus if we choose $\delta = r/2$ it is impossible to find an a in $[0, 1]$ with $|x - a| < \delta$, and so x is not a point of closure of $[0, 1]$. The only restriction we placed on x was that x should be greater than 1. Thus we may conclude that no point x with $x > 1$ is a point of closure of $[0, 1]$. A similar argument shows no point x with $x < 0$ is a point of closure of $[0, 1]$ and so $\overline{[0, 1]} = [0, 1]$, i.e. $[0, 1]$ is closed.

Exercises

1. Let $\mathbb{R}$ have the usual norm. Which of the following sets are (a) open? (b) closed?

$$\{2\}, \quad [2, 3], \quad \{t : t^2 < 1\}, \quad \{2, 3\}, \quad \{t : 0 < t \leq 4\}.$$

2. Show that

$$\mathrm{int}(\mathrm{int}\,A) = \mathrm{int}\,A \quad \text{and} \quad \mathrm{int}(A \cap B) = \mathrm{int}\,A \cap \mathrm{int}\,B.$$

Give an example to show that $\mathrm{int}(A \cup B) = \mathrm{int}\,A \cup \mathrm{int}\,B$ is not necessarily true.

3. Show that if $A \subseteq B$ then $\overline{A} \subseteq \overline{B}$. (In particular, this shows that if B is a closed set and $A \subseteq B$ then $\overline{A} \subseteq B$.)

4. The set $\{x : \|x - a\| \leq r\}$ is closed in a normed linear space X.

5. The set $\{x\}$, which consists of a single point in the normed linear space X, is closed.

6. The set $[0, 1)$ is neither open nor closed in $(\mathbb{R}, |\cdot|)$. (This shows that the ideas of open and closed are not mutually exclusive – there are some sets in each normed linear space which are neither open nor closed.)

7. The set $[0,1]$ is closed in $\mathbb{R}$. The set $\{(s,0) : 0 \leq s \leq 1\}$ is not closed in $\mathbb{R}^2$. (Thus the 'place of abode' of the set affects the question of its closure.)

There now follow several results which elucidate the properties of open and closed sets.

Lemma 3.3 *A set A in a normed linear space X is closed if and only if $X \setminus A$ is open.*

Proof. Suppose A is closed in X. Take x in $X \setminus A$. Then x cannot be a point of closure of A, otherwise A would not be closed. Hence, there must exist some $\delta > 0$ with no a in A such that $\|x - a\| < \delta$. Thus $B_\delta(x) \subseteq X \setminus A$ and so $X \setminus A$ is open. On the other hand, suppose $X \setminus A$ is open. Take x in $X \setminus A$ again. Then there is some value $r > 0$ such that $B_r(x)$ lies in $X \setminus A$. Hence $B_r(x)$ does not intersect A and so x cannot be a point of closure of A. Since the only restriction we placed on x was that x lay in $X \setminus A$, no point of $X \setminus A$ is a point of closure of A. Thus every point of closure of A lies in A and so A is closed. ∎

There is one strange anomaly of the definitions which we now mention. A particular subset of the normed linear space X is X itself. It is a triviality that for each x in the set X there is an open ball $B_r(x)$ which lies inside the set X. Hence the set X is open when considered as a subset of itself. However, all the points of closure of any set A in X are themselves defined to be points in X, so for any set $A \subseteq X$ we have $\overline{A} \subseteq X$. If $X = A$ we have that $\overline{X} \subseteq X$ and so X is closed. The conclusion is that the set X is both open and closed in the normed linear space X. A consequence of **3.3** is that the empty set is also an open and closed subset of X.

Lemma 3.4 *The union of any family of open sets is open. The intersection of any family of closed sets is closed.*

Proof. Let $\mathcal{G}$ be a family of open sets in a normed linear space X. We write their union as $\bigcup_{G \in \mathcal{G}} G$. Take $g \in \bigcup_{G \in \mathcal{G}} G$. Then $g \in G$ for some G in the family, and since G is open there is a ball $B_r(g)$, $r > 0$ such that $B_r(g) \subseteq G$. Hence $B_r(g) \subseteq \bigcup_{G \in \mathcal{G}} G$ and so $\bigcup_{G \in \mathcal{G}} G$ is open. Let $\mathcal{F}$ be a family of closed sets in X. Then the family $\mathcal{G}$

of complements of members of $\mathcal{F}$ is a family of open sets and so $\mathcal{G}$ is open. Now $\bigcap_{F \in \mathcal{F}} F$ can be written as $X \setminus \bigcup_{G \in \mathcal{G}} G$. It follows from Lemma **3.3** that $\bigcap_{F \in \mathcal{F}} F$ is closed. ∎

Lemma 3.5 *The intersection of finitely many open sets is open. The union of finitely many closed sets is closed.*

Proof. Let $G_1, G_2, ..., G_n$ be open sets in a normed linear space X, and set $\mathcal{G} = G_1 \cap G_2 \cap ... \cap G_n$. Take x in $\mathcal{G}$. Since each G_i is open, $1 \leq i \leq n$, there exists an $r_i > 0$ such that $B_{r_i}(x) \subseteq G_i$. Set $r = \min_i r_i$. Then $B_r(x) \subseteq G_i$ for all i, $1 \leq i \leq n$, and so $B_r(x) \subseteq G$. Hence G is open. The second half of the result is proved by a complement argument. ∎

Note that it is impossible to remove the condition of finiteness from **3.5**. The intersection of an infinite collection of open sets may be open or closed as the Exercises below show.

Exercises

1. Prove that $\mathbb{Z}$ (the integers) is a closed subset of $\mathbb{R}$.

2. If $A = \{x : \|x\| < 1\}$ and $B = \{x : \|x\| \leq 1\}$ show that $\overline{A} = B$.

3. Prove that $\overline{A \cup B} = \overline{A} \cup \overline{B}$ for any pair of sets A, B in a normed linear space X.

4. Is $\overline{A \cap B} = \overline{A} \cap \overline{B}$ for all pairs of sets A and B in a normed linear space?

5. Suppose f is a continuous function from a normed linear space X into a normed linear space Y. If $A \subseteq X$ show that $f(\overline{A}) \subseteq \overline{f(A)}$.

6. Show that $\bigcap_{n=1}^{\infty}(-\frac{1}{n}, 1 + \frac{1}{n})$ is closed in $\mathbb{R}$.

7. Show that $\bigcup_{n=2}^{\infty}[\frac{1}{n}, 1 - \frac{1}{n}]$ is open in $\mathbb{R}$.

8. Let f be a continuous, real-valued function on $\mathbb{R}$. Show that the set $\{(a, b) : b \leq f(a)\}$ is closed in $\mathbb{R}^2$.

9. Let f and g be continuous mappings from a normed linear space X into a second normed linear space Y. Prove that $\{x : f(x) = g(x)\}$ is closed in X.

10. Show that $\{(a, \frac{1}{a}) : a > 0\}$ is closed in $\mathbb{R}^2$.

Open and closed sets can be used to give an alternative definition of the continuity of a function. This new viewpoint is potentially very attractive, because it removes any mention of the quantifiers ϵ and δ, which on first encounter prove to be somewhat awkward to work with. However, we will find that the new concepts offer more or less the same degree of difficulty. Nevertheless, the more ways we have of expressing a property, the greater will be our flexibility in seeking to establish that a certain structure (in this case a function) has that given property.

Definition 3.6 *Let f be a mapping from a normed linear space X into a normed linear space Y. If A is a set in X and B is a set in Y, then we define*

$$f(A) = \{f(a) : a \in A\},$$

$$f^{-1}(B) = \{x \in X : f(x) \in B\}.$$

Notice that in this definition, f is *not* presumed to be invertible, or even one-to-one. The notation $f^{-1}(B)$ has nothing to do with whether f has an inverse or not. This makes the use of the superscript -1 rather confusing, since it is open now to at least three interpretations. If we write $[f(3)]^{-1}$, we are referring to the number which is the multiplicative inverse of the number $f(3)$. If we encounter the notation $f^{-1}(3)$ for a function from $\mathbb{R}$ to $\mathbb{R}$, then we would interpret this as implying that f is an invertible function and the notation as referring to the inverse image of 3 under the function f. For the same function the notation $f^{-1}([0,1])$ is *not* to be interpreted as implying anything about invertibility of f. It simply stands for the set of real numbers which are mapped into $[0,1]$ by f. If we want to denote the set of real numbers which are mapped onto the point 3 by a (not necessarily invertible or one-to-one) function f then we would write $f^{-1}(\{3\})$ to distinguish it from $f^{-1}(3)$. Figure 3.3 illustrates one of these sets.

The next theorem describes two alternatives to definition 2.8. One of them has already been encountered in 2.10 and is restated here for completeness only.

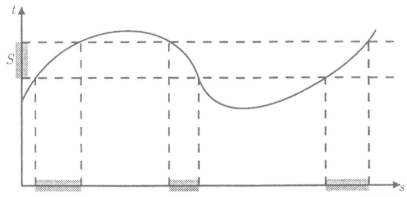

The union of the shaded areas on the s-axis represents $f^{-1}(S)$

Figure 3.3: An example of a set of the form $f^{-1}(S)$.

Theorem 3.7 *Let f be a mapping from one normed linear space X into another, Y. Then the following are equivalent:*
(i) f is continuous;
(ii) whenever $\{x_n\}$ is a sequence in X with limit x, $\{f(x_n)\}$ is a sequence with limit $f(x)$ in Y;
(iii) for every open set $G \subseteq Y$, $f^{-1}(G)$ is open in X;
(iv) for every closed set F in Y, $f^{-1}(F)$ is closed in X.

Proof. The substance of **2.10** is that statements (i) and (ii) are equivalent. We will now show that (i) and (iii) are equivalent.
(i)$\Rightarrow$(iii) Let G be an open set in Y. It must be shown that $f^{-1}(G)$ is open in X. As usual this is done by taking a general point x_0 in $f^{-1}(G)$ and showing that it can be surrounded by an open ball (of carefully chosen radius!) which also lies in $f^{-1}(G)$. Set $y_0 = f(x_0)$. Since $y_0 \in G$ there is some $r > 0$ such that the ball $B_r(y_0)$ lies in G, i.e., $\{y \in Y : \|y - y_0\| < r\} \subseteq G$. Since f is continuous, there exists a $\delta > 0$ such that whenever $\|x - x_0\| < \delta$, then $\|f(x) - f(x_0)\| < r$. Now the ball $B_\delta(x_0)$ is our candidate for an open ball surrounding x_0 and lying in $f^{-1}(G)$. That this is indeed the case follows from the observation that $f(B_\delta(x_0)) \subseteq B_r(y_0) \subseteq G$.
(iii)$\Rightarrow$(i) We have to establish using **2.8** that f is continuous. Any proof of this nature must establish the truth of a certain condition for all values of $\epsilon > 0$ and for all points x in X. Thus the correct

way to start the proof is to suppose x_0 is in X and ϵ is a real, positive number. Now we must manufacture the δ which ensures that whenever $\|x - x_0\| < \delta$ this guarantees $\|f(x) - f(x_0)\| < \epsilon$. Set $G = B_\epsilon(f(x_0))$. Then G is an open set in Y and so by hypothesis $f^{-1}(G)$ is open in X. Also x_0 is in $f^{-1}(G)$. Hence there is a ball $B_\delta(x_0)$ which is contained in $f^{-1}(G)$ for some suitable $\delta > 0$. Rewriting this as $\{x : \|x - x_0\| < \delta\} \subseteq f^{-1}(G)$ and recalling what $f^{-1}(G)$ means makes it plain that $f(x)$ is in G whenever x is in $B_\delta(x_0)$. Again, simply by rewriting this symbolically, we see that $\|f(x) - f(x_0)\| < \epsilon$ whenever $\|x - x_0\| < \delta$ (which is what it means to say x is in $B_\delta(x_0)$). Thus f is indeed continuous. We could now show that (i) and (iv) are equivalent by arguments modelled on but different from those above. This would be an instructive exercise, but by now the idea of using set complementation should spring easily to mind. The starting point is to observe that for any set G in Y, $f^{-1}(Y \setminus G) = X \setminus f^{-1}(G)$. The details are left to the reader. ∎

A whole host of simple corollaries follow from this result. Some are given in the next set of exercises. They are therefore not to be missed – they will enlarge your understanding of what can be achieved using **3.7**. An important observation which we could have made at an earlier stage but chose to delay is the following.

Corollary 3.8 *Let A be a subset of the normed linear space X. Then the closure of A is a closed subset of X.*

Proof. This proof involves no calculation at all, but rather a set of observations on closure and continuity. One of the ways of describing $\overline{A}$ is to write $\overline{A} = \{x \in X : f(x) = 0\}$ where $f : X \to \mathbb{R}$ is the distance function $f(x) = \text{dist}(x, A)$. By 2.13, f is continuous and since $\overline{A} = f^{-1}(\{0\})$ and $\{0\}$ is closed in $\mathbb{R}$, it follows from **3.7** that $\overline{A}$ is closed. ∎

We saw at an early stage that it was possible to rephrase continuity in terms of sequential ideas. This is also possible with the idea of closure.

Lemma 3.9 *Let A be a subset of the normed linear space X.*
(i) A point x in X is in $\overline{A}$ if and only if x is the limit of some sequence of points in A.
(ii) The set A is closed if and only if, whenever $\{a_n\}$ is a sequence of points in A converging to $x \in X$, then x is also in A.

Proof. Let us observe at the outset that (ii) is simply a combination of (i) and the definition of $\overline{A}$. Therefore, we only have something to prove in (i), the second statement being given because that is how the lemma is most often used. Suppose that x is the limit of the sequence $\{a_n\}$ whose points lie in A. Then for any $\epsilon > 0$ there is an N such that $\|x - a_N\| < \epsilon$. This means that x is a point of closure of A. Now suppose $x \in \overline{A}$. Then, in particular, for each value of n in $\mathbb{N}$ there is an $a_n \in A$ with $\|x - a_n\| < 1/n$. Hence the sequence $\{a_n\}$ converges to x. ∎

Exercises

1. Let f be a continuous mapping from a normed linear space X into $\mathbb{R}$. Then the following sets are open:

$$\{x : f(x) < \alpha\}, \quad \{x : f(x) > \beta\}, \quad \{x : \alpha < f(x) < \beta\}.$$

   The following sets are closed

$$\{x : f(x) \le \alpha\}, \quad \{x : f(x) \ge \beta\}, \quad \{x : \alpha \le f(x) \le \beta\}.$$

   What can you say about the set $\{x : \alpha < f(x) \le \beta\}$?

2. Obtain short proofs based on **3.7** for questions 8 and 9 in the previous set of exercises.

3. Show that $\{(s,t) : s^2 + t^2 < 1\}$ is open in $\mathbb{R}^2$ with any of the three norms $\| \cdot \|_1, \| \cdot \|_2$ or $\| \cdot \|_\infty$.

4. Show that in $\mathbb{R}^3$,
   (i) the set $\{(s,t,u) : s,t,u > 0\}$ is open
   (ii) the set $\{(s,t,u) : s = t = u\}$ is not open.

5. Show that if G is open in the normed linear space X and A is a set with $A \cap G$ empty, then $\overline{A} \cap G$ is also empty.

6. Let A be a subset of the normed linear space X satisfying $\lambda a \in A$ whenever $a \in A$ and $\lambda \geq 0$. Show that A is closed if and only if $A \cap \{x : \|x\| \leq 1\}$ is closed.

7. For any set A in a normed linear space X, show that $\text{dist}(x, A) = \text{dist}(x, \overline{A})$ for all x in X.

8. Show that $\{(s,0) : s \in \mathbb{R}\}$ is closed in $\mathbb{R}^2$ with any of the norms $\|\cdot\|_1$, $\|\cdot\|_2$ or $\|\cdot\|_\infty$.

9. Define the distance between two sets A and B in a normed linear space X as

$$d(A, B) = \inf\{\|a - b\| : a \in A, \, b \in B\}.$$

Is it possible to have two closed sets A and B in X such that $d(A, B) = 0$ and $A \cap B$ empty? It might help to consider $X = \mathbb{R}^2$,

$$A = \{(s,0) : s \in \mathbb{R}\} \quad \text{and} \quad B = \{(s, e^{-s^2}) : s \in \mathbb{R}\}.$$

10. Let A and B be disjoint closed sets in a normed linear space X. Construct a continuous function $f : X \to [0,1]$ such that $f(a) = 0$ for all a in A, and $f(b) = 1$ for all b in B.

11. Let E be a linear subspace of the normed linear space X. Suppose that for some $\delta > 0$, E contains all x in X with $\|x\| < \delta$. Show that $E = X$. Deduce that if $E \neq X$, then E has no interior points.

12. Give an example of a set A in $\mathbb{R}^2$ such that $\overline{A} = \mathbb{R}^2$ and $\text{int} A$ is empty.

13. Let $(X, \|\cdot\|_X)$ and $(Y, \|\cdot\|_Y)$ be normed linear spaces. Set $Z = (X \times Y, \|\cdot\|_Z)$, where as usual $X \times Y$ is the set of ordered pairs (x, y), where x lies in X and y in Y. The norm is defined by

$$\|z\|_Z = \|(x,y)\|_Z = \max\{\|x\|_X, \|y\|_Y\}.$$

For sets $A \in X$ and $B \in Y$ show that
(i) $A \times B$ is open if and only if A and B are open
(ii) $A \times B$ is closed if and only if A and B are closed.

14. The following constructs the Cantor set in $\mathbb{R}$: begin with the unit interval $[0,1]$. The set F_1 is obtained from this set by removing the middle third, so that

$$F_1 = \left[0, \frac{1}{3}\right] \cup \left[\frac{2}{3}, 1\right].$$

Now F_2 is obtained from F_1 by removing the middle third from each of the constituent intervals of F_1. Hence

$$F_2 = \left[0, \frac{1}{9}\right] \cup \left[\frac{2}{9}, \frac{1}{3}\right] \cup \left[\frac{2}{3}, \frac{7}{9}\right] \cup \left[\frac{8}{9}, 1\right].$$

In general, F_n is the union of 2^n intervals, each of which has the form

$$\left[\frac{k}{3^n}, \frac{k+1}{3^n}\right],$$

where k lies between 0 and 3^n. The Cantor set F is what remains after this process has been carried out for all n in $\mathbb{N}$. Show that

(i) F is closed in $\mathbb{R}$

(ii) $\mathrm{int}\, F$ is empty

(iii) F contains no non-empty open set.

(iv) the complement of F can be expressed as a countable union of open intervals. (If you are unfamiliar with the definition of a countable set, then Appendix A gives a brief introduction to this concept.)

4

Denseness, separability and completeness

At first sight the structure of a given normed linear space may seem quite complicated. For example, the space $C([0,1], \|.\|_\infty)$ consists of *all* continuous, real-valued functions on $[0,1]$. This includes functions whose behaviour is so wild that, although they are continuous at all points on $[0,1]$, they are differentiable nowhere on that interval. We cannot begin to draw the graph of such a function, but a function with this property is constructed in chapter 5. We might ask whether such functions are intrinsically very different from ones which are differentiable everywhere on $[0,1]$. If they are, then they make $C[0,1]$ more mysterious in that it not only consists of functions whose graphs we can visualise, but also a large and important group of functions about which we have very little intuitive or visual knowledge. Before we can address such a question we will have to decide what we mean by the phrase 'intrinsically different' which at the moment is woefully vague.

One of the very common ways to explore the structure of a normed linear space is to look for sets whose closure is the whole space.

Definition 4.1 *A subset A of the normed linear space X is said to be dense in X if $\overline{A} = X$.*

If a set A is dense in X, then the fact that $\overline{A} = X$ means every point in X is a point of closure of A. Thus given any x in X and a 'tolerance' $\epsilon > 0$, there is a point a in A such that $\|x - a\| < \epsilon$. Sometimes we say that any element of X may be approximated *arbitrarily closely* by elements from the set A. We shall see very soon (and in fact could see now from **4.1**) that the rationals are dense in the reals. This is a fact on which scientists rely every day. If they are involved in a computation which demands the use of $\sqrt{2}$, then the fact that this is an irrational number, with a non-recurring decimal expansion, which consequently will fit in no computing machine designed to date, does not deter them. They decide what sort of tolerance is appropriate for their calculation and replace $\sqrt{2}$ by a rational value which is within the required tolerance of $\sqrt{2}$. For example if the tolerance were 0.001 then the scientist might well use the value 1.414 as a suitable rational replacement for $\sqrt{2}$, because $|\sqrt{2} - 1.414| < 0.001$.

This is broadly the same route that the analyst follows using density. He may have a seemingly large normed linear space which includes some rather mysterious looking and hard to handle elements. He then looks for a set A which is dense in X and has as simple a structure as possible. The second criterion here, that of simple structure, is vital. It is easy to manufacture dense sets for a given normed linear space (for example the set $\mathbb{R} \setminus \{0\}$ is dense in $\mathbb{R}$), but the skill lies in making A as small and as simple as possible.

Lemma 4.2 *Let X be a normed linear space. Then the following statements are equivalent:*
(i) the set A is dense in X,
(ii) the set $X \setminus A$ has no interior point.

Proof. $(i) \Rightarrow (ii)$ We use the technique of contrapositive argument here. We show that if $X \setminus A$ has an interior point then $\overline{A} \neq X$. If x is an interior point of $X \setminus A$, then there is a ball $B_r(x)$ with $r > 0$ which is contained completely in $X \setminus A$. It is impossible for this point x to lie in $\overline{A}$ and so $\overline{A} \neq X$.
$(ii) \Rightarrow (i)$ If x is a point in X, then it is not an interior point of $X \setminus A$ and so every ball $B_r(x)$ with $r > 0$ intersects A. Thus x is a point in $\overline{A}$ and so $\overline{A} = X$. ∎

Corollary 4.3 *The rational numbers are dense in the real numbers.*

Proof. This result is an immediate observation from **4.2**. ∎

We now give an example of how the property of denseness can be passed from one normed linear space to another.

Lemma 4.4 *Let f be a continuous mapping from the normed linear space X onto the normed linear space Y. If A is dense in X, then $f(A)$ is dense in Y.*

Proof. Take a point y_0 in Y and an $\epsilon > 0$. We must show that there is a point y in $f(A)$ such that $\|y_0 - y\| < \epsilon$. This must be achieved by looking 'back' to information about A in X. Since f maps X *onto* Y, there exists a point x_0 in X such that $f(x_0) = y_0$. Since f is continuous at x_0, there is a $\delta > 0$ such that if $\|x - x_0\| < \delta$, then $\|f(x) - f(x_0)\| < \epsilon$. Since $\overline{A} = X$, the set $\{x \in X : \|x - x_0\| < \delta\}$ must contain at least one element a in A. Consequently, $\|f(a) - y_0\| < \epsilon$ as required. ∎

We have already pointed out that for a dense set A to simplify our understanding of a given normed linear space X, the set A must be simple in some sense. One of the ways it can be simple is for it to be countable. (Appendix A contains a simple exposition of the idea of countability.)

Definition 4.5 *A normed linear space is said to be separable if it contains a countable dense subset.*

The chief advantage of a countable set is that we can decide which element is the first element, which is the second, and so on. Thus if A is a countable set we can write A as

$$A = \{a_1, a_2, a_3, \ldots\} = \{a_n\}_1^\infty.$$

If such a set is dense in a normed linear space X, then given any x in X and a tolerance $\epsilon > 0$, there is a value $n \in \mathbb{N}$ such that $\|x - a_n\| < \epsilon$. From our earlier discussions it is now clear that $\mathbb{R}$ is a separable normed linear space because it contains the countable dense set $\mathbb{Q}$ – the rationals.

Lemma 4.6 *If f is a continuous mapping of a separable normed linear space X onto a normed linear space Y, then Y is separable.*

Proof. If A is a countable, dense set in X, then $f(A)$ is a countable set in Y (Appendix A). Furthermore, $f(A)$ is dense in Y by **4.4**. Hence Y is separable. ∎

Lemma 4.7 *Every subspace of a separable normed linear space is separable.*

Proof. Let X be a separable normed linear space, and $\{x_n\}$ a countable dense set in X. Let E be a subspace of X. Define points t_{mn}, $1 \leq m, n < \infty$, by choosing t_{mn} in $E \cap B_{1/m}(x_n)$ whenever this set is non-empty and $t_{mn} = \theta$ otherwise. Then the set $\{t_{mn} : 1 \leq m, n < \infty\}$ is certainly countable (see Appendix A). It is also a subset of E which is dense in E as the following computation will show. Choose x in E and $\epsilon > 0$. Pick $m \in \mathbb{N}$ such that $2/m \leq \epsilon$. Since $\{x_n\}$ is dense X, there is a value of n in $\mathbb{N}$ such that $\|x - x_n\| < 1/m$. Consequently, the set $B_{1/m}(x_n) \cap E$ is non-empty and so there is a point t_{mn} in this set. Finally,

$$
\begin{aligned}
\|x - t_{mn}\| &= \|x - x_n + x_n - t_{mn}\| \\
&\leq \|x - x_n\| + \|x_n - t_{mn}\| \\
&< \frac{1}{m} + \frac{1}{m} \\
&= \frac{2}{m} \\
&\leq \epsilon.
\end{aligned}
$$

This shows that $\{t_{mn} : 1 \leq m, n < \infty\}$ is dense in E. ∎

Notice that in this proof the t_{mn} which were set equal to θ never played a role. This was just a notational convenience so that each t_{mn} was defined whatever the values of m and n were. If we had not done this then the set $\{t_{mn} : 1 \leq m, n < \infty\}$ would have been poorly defined. If m and n were such that $E \cap B_{1/m}(x_n)$ were by chance empty then the reader would not know what to take for t_{mn}.

We conclude this discussion of separability by giving a criterion which determines when a space is not separable.

Lemma 4.8 *Let E be an uncountable subset of the normed linear space X, with the property that there is an $\epsilon > 0$ such that whenever g and h are distinct members of E we have $\|g - h\| \geq \epsilon$. Then X is not separable.*

Proof. Let A be a dense subset of X. For each g in E, choose a_g in A so that $\|g - a_g\| < \frac{1}{2}\epsilon$. If g and h are distinct points in E, then a_g

and a_h are distinct points in A, otherwise we would have

$$
\begin{aligned}
\|g - h\| &= \|g - a_g + a_g - h\| \\
&\leq \|g - a_g\| + \|a_g - h\| \\
&= \|g - a_g\| + \|a_h - h\| \\
&< \epsilon/2 + \epsilon/2 \\
&= \epsilon.
\end{aligned}
$$

This contradicts the fact that points in E are separated by at least ϵ. Now the force of the fact that the distinctness of g and h in E implies the distinctness of a_g and a_h in A is that there is a one-to-one correspondence between E and A. Hence A is uncountable, and so it is impossible for X to be separable. ∎

Now we come to our last desirable property for a normed linear space in this chapter – that of completeness.

Definition 4.9 *A sequence $\{x_n\}$ in a normed linear space is said to be Cauchy if given $\epsilon > 0$ there exists an N such that $\|x_p - x_q\| \leq \epsilon$ whenever $p, q \geq N$.*

Roughly speaking, Cauchy sequences are ones in which all 'far out' members are squeezed close together. Every convergent sequence is necessarily a Cauchy sequence. To see this suppose $\{x_n\}$ is a sequence in the normed linear space with $x_n \to x$. Take $\epsilon > 0$. We must establish that there is some value of N such that $\|x_p - x_q\| \leq \epsilon$ whenever $p, q \geq N$. Since $\{x_n\}$ is convergent to x, we may choose N such that $\|x_n - x\| \leq \epsilon/2$ for all $n \geq N$. Then if $p, q \geq N$, we have

$$
\|x_p - x_q\| = \|x_p - x + x - x_q\| \leq \|x_p - x\| + \|x - x_q\| \leq \frac{\epsilon}{2} + \frac{\epsilon}{2} = \epsilon.
$$

Thus the sequence is Cauchy. We naturally expect that the reverse is true. If the 'later' members of a sequence are squeezed up into successively smaller balls as we progress 'outwards' in the sequence then surely the sequence must be converging. This is not always the case.

Definition 4.10 *A set A in the normed linear space X is complete if every Cauchy sequence in A has limit in A.*

It can be seen that completeness will have something to do with closure. For example, the set $(0,1)$ is *not* complete in $\mathbb{R}$. The sequence $\{1/n\}_{n=2}^{\infty}$ is a convergent sequence with limit 0. Hence it is a Cauchy sequence in $(0,1)$, but because its limit lies outside $(0,1)$, this set is not complete. This is always the cause of a set being incomplete in practice – the desired limit points of certain Cauchy sequences have somehow been omitted from the set. Since closure and completeness are closely linked, we usually only refer to completeness in the case of a normed linear space X. Within this context, it is such an important concept that a complete normed linear space is given a special name. It is called a *Banach space*, after the great Polish mathematician S. Banach, whose research did so much to establish and advance the progress of analysis in normed linear spaces. The connection between closure and completeness is laid bare in the next result.

Lemma 4.11 *Let X be a normed linear space.*
(i) If X is complete, then every closed subset of X is also complete.
(ii) If A is contained in X and A is complete, then A is closed.
(ii) If X is complete and $A \subset X$, then A is complete if and only if A is closed.

Proof. (i) Let A be a closed subset of X, and let $\{a_n\}$ be a Cauchy sequence in A. Since X is complete, $\{a_n\}$ converges to some point x in X. Since A is closed, $x \in A$ and so A is complete.
(ii) Take a point x in $\overline{A}$. By **3.9**, there exists a sequence $\{a_n\}$ contained in A, such that $a_n \to x$. Since A is complete, the Cauchy sequence $\{a_n\}$ must converge to a point in A. That is, $x \in A$. Thus we conclude A is closed.
(iii) This is a consequence of (i) and (ii). ∎

If $\{x_n\}$ is a sequence of points in a normed linear space then a subsequence is an 'increasing selection' of members of $\{x_n\}$. We often write a subsequence as $\{x_{n_k}; k = 1, 2, \ldots,$ where $n_1 < n_2 < \ldots\}$ or more simply as $\{x_{n_k}\}_{k=1}^{\infty}$. For example, $x_1, x_4, x_5, x_{23}, x_{36}, \ldots$ is a subsequence of $x_1, x_2, \ldots$. However, $x_1, x_4, x_4, x_3, x_6, x_7, \ldots$ is not, because the point x_4 is not permitted to be selected twice, neither is the point x_3 permitted to be selected *after* the point x_4. Both selections violate the stipulation that the selection be increasing.

Lemma 4.12 *If $\{x_n\}$ is a Cauchy sequence in a normed linear space X, and some subsequence $\{x_{n_k}\}$ converges to x in X, then $\{x_n\}$ itself converges to x.*

Proof. We start in the usual manner. Take $\epsilon > 0$. Since $\{x_n\}$ is Cauchy there is an N such that $\|x_p - x_q\| \leq \frac{1}{2}\epsilon$ whenever $p, q \geq N$. Since $\{x_{n_k}\}$ converges to x, there is an m such that $n_m \geq N$ and $\|x_{n_m} - x\| \leq \frac{1}{2}\epsilon$. Now applying the first statement with $q = n_m$ gives $\|x_p - x_{n_m}\| \leq \frac{1}{2}\epsilon$ for all $p \geq N$. Finally, we compute as follows for all $p \geq N$:

$$\|x_p - x\| \leq \|x_p - x_{n_m}\| + \|x_{n_m} - x\| \leq \frac{1}{2}\epsilon + \frac{1}{2}\epsilon = \epsilon.$$

Hence $\{x_n\}$ converges to x. ∎

At last we now establish that some complete normed linear spaces exist!

Theorem 4.13 *The linear space $\mathbb{R}^n$ is complete with any of the three norms $\|\cdot\|_1$, $\|\cdot\|_2$ or $\|\cdot\|_\infty$.*

Proof. We use $\|\cdot\|_\infty$ and adopt the notation x_k for the kth member of a Cauchy sequence in $\mathbb{R}^n$. This k^{th} member will have coordinates

$$x_k(1), x_k(2), \ldots, x_k(n).$$

We shall also take as a proven fact that a Cauchy sequence of real numbers converges to a real number. There is an opportunity in the exercises to recap on this! So suppose $\{x_k\}$ is a Cauchy sequence in $\mathbb{R}^n$. Take $\epsilon > 0$. Then there exists an N in $\mathbb{N}$ such that whenever $p, q \geq N$, then $\|x_p - x_q\|_\infty \leq \epsilon$. For any fixed $i, 1 \leq i \leq n$, this implies that $|x_p(i) - x_q(i)| \leq \epsilon$ for all $p, q \geq N$. Hence $\{x_k(i)\}$ is a Cauchy sequence of real numbers for each $1 \leq i \leq n$, and so converges to a limit, which we call $x(i)$. In this way we define the element x in $\mathbb{R}^n$, and as $x_k(i) \to x(i)$, $1 \leq i \leq n$, we have $x_k \to x_0$. Hence any Cauchy sequence in $\mathbb{R}^n$ is convergent to a point in $\mathbb{R}^n$, as required. ∎

We will conclude this section with an important theorem known as Cantor's intersection theorem. This result will be used in later chapters, but belongs naturally in this section on completeness. We need some terminology before we can introduce the theorem.

Definition 4.14 *A sequence of sets $\{F_n\}_1^\infty$ is said to be decreasing if $F_{n+1} \subseteq F_n$ for $n = 1, 2, \ldots$. The sequence is said to be increasing if $F_{n+1} \supseteq F_n$ for $n = 1, 2, \ldots$. The diameter of a set A in a normed linear space X is the number $\sup\{\|a - b\| : a, b \in A\}$, whenever this is finite. If the set $\{\|a - b\| : a, b \in A\}$ is unbounded we say that the diameter of A, written $\mathrm{diam}(A)$, is ∞.*

Theorem 4.15 *(Cantor's Intersection Theorem) Let A be a complete subset of a normed linear space X. Let $\{F_n\}$ be a decreasing sequence of non-empty, closed subsets of A such that $\mathrm{diam}(F_n) \to 0$ as $n \to \infty$. Then $\bigcap_{n=1}^\infty F_n$ contains exactly one point.*

Proof. Firstly, there is at most one point in $\bigcap_{n=1}^\infty F_n$, since if $a \neq b$ and $a, b \in \bigcap_{n=1}^\infty F_n$ then $a, b \in F_n$ for each n, and so $\mathrm{diam}(F_n) \geq \|a - b\| > 0$. This would contradict the assumption that $\mathrm{diam}(F_n) \to 0$. Now we must show that there *is* a point in $\bigcap_{n=1}^\infty F_n$. For each n, choose a point x_n in F_n, so obtaining a sequence $\{x_n\}$. If $p, q \geq n$ then x_p and x_q are each in F_n and so $\|x_p - x_q\| \leq \mathrm{diam}(F_n)$. Thus, given $\epsilon > 0$, taking n sufficiently large to ensure $\mathrm{diam}(F_n) < \epsilon$ will ensure $\|x_p - x_q\| < \epsilon$ for all $p, q \geq n$. This means that $\{x_n\}$ is a Cauchy sequence. Since A is complete, $x_n \to x$ where $x \in A$. This point x is our candidate for the point in $\bigcap_{n=1}^\infty F_n$. That it actually does the job is seen as follows. For a fixed value of p we have that each $x_n \in F_p$ for $n \geq p$. Since $x_n \to x$ and F_p is closed, we must have $x \in F_p$ by **3.9**. Hence $x \in \bigcap_{n=1}^\infty F_n$ as required. ∎

Exercises

1. Let $B(S)$ denote the bounded functions on $S = [0, 1]$. Let F denote the set of functions f for which there exists a sequence of sets $S_1, \ldots, S_k$ with f constant on each set S_i, $1 \leq i \leq k$. Prove F is a dense subset of $B(S)$.

2. Let G_1 be an open, dense set in a normed linear space X. Prove that for any $x \in X$ and $r > 0$, the set $B_r(x) \cap G_1$ contains a ball. Deduce that if G_1 and G_2 are open and dense in X, then $G_1 \cap G_2$ is also open and dense in X.

3. Give an example to show that the condition $\{\|x_n - x_{n+1}\|\} \to 0$ is *not* sufficient for $\{x_n\}$ to be a Cauchy sequence in a normed linear space.

4. The following guides you through the construction of the proof of the fact that a Cauchy sequence of real numbers converges. Let $\{a_n\}$ be a sequence of real numbers.
(i) Show that $\{a_n\}$ contains a subsequence which is either non-decreasing or non-increasing.
(ii) Use (i) to show that every bounded sequence has a convergent subsequence.
Now let $\{a_n\}$ be a Cauchy sequence of real numbers.
(iii) Show that $\{a_n\}$ is bounded.
(iv) Deduce that $\{a_n\}$ contains a convergent subsequence.
(v) Show that $\{a_n\}$ converges.

5. The following shows how to determine that the linear space $C[0,1]$ with the norm

$$\|x\| = \int_0^1 |x(s)| ds$$

is incomplete.
(a) Define

$$x_n(s) = \begin{cases} -1 & s \in [-1, -\frac{1}{n}] \\ ns & s \in (-\frac{1}{n}, \frac{1}{n}) \\ 1 & s \in [\frac{1}{n}, 1] \end{cases}$$

Sketch x_n and compute $\|x_p - x_q\|$ for arbitrary values of p and q.
(b) Show that $\{x_n\}$ is a Cauchy sequence.
(c) Define x as the function

$$x(s) = \begin{cases} -1 & s \in [-1, 0) \\ 1 & s \in [0, 1]. \end{cases}$$

Compute $\|x - x_n\|$ for abitrary values of n in $\mathbb{N}$.
(d) Show that $\|x_n - x\| \to 0$ as $n \to \infty$.
(e) Show that $C[0,1]$ with the given norm is not complete.

6. Show X is a separable normed linear space if and only for each $\epsilon > 0$ there exists a countable collection of sets of diameter at most ϵ such that X is continued in their union.

7. Let $\|\cdot\|_1$ and $\|\cdot\|_2$ be two norms defined on a linear space X. Suppose there exist constants $a > 0$ and $b > 0$ such that

$$a\|x\|_1 \le \|x\|_2 \le b\|x\|_1, \quad \text{for all } x \text{ in } X.$$

Show that $(X, \| \cdot \|_1)$ is complete if and only if $(X, \| \cdot \|_2)$ is complete.

8. Let X be a complete normed linear space. Let $A_1, A_2, \ldots$ be a sequence of closed sets in X with $\bigcup_{n=1}^{\infty} A_n = X$. Show that $\bigcup_n (\text{int} A_n)$ is dense in X.

9. Let X be complete, and $\{F_n\}$ a sequence of continuous functions from X to $\mathbb{R}$ with the property that $f_n(x) \to f(x)$ as $n \to \infty$ for all $x \in X$. Prove that the set of points in X at which f is continuous is dense in X. [Hint: let

$$A_{mn} = \{x : |f_p(x) - f_n(x)| \leq 1/m \text{ for all } p \geq n\}.$$

Show using the previous question that $G_m = \bigcup_n \text{int} A_{mn}$ is dense in X, and prove that f is continuous at each point of $\bigcap_m G_m$.]

5

Function spaces

Our claim throughout this book is that the language of abstract analysis is a useful tool in concrete situations. So we turn aside for a while from our development of the mainstream of abstract ideas and concentrate in this chapter on a practical example of a normed linear space. This is quite a lengthy chapter, and in it we shall discover that some of the ideas we have been making precise in our earlier discussion turn out to be quite difficult to check in a given situation.

We shall consider a set S in the real numbers $\mathbb{R}$. Very often S will be a simple set, for example $[0,1]$ or $(-1,1)$, but we don't want to restrict ourselves to that sort of choice just yet. We have already encountered the linear spaces $B(S)$ and $C(S)$. Recall that $B(S)$ consists of all bounded, real-valued functions on S, and that $C(S)$ is the subspace of $B(S)$ consisting of all bounded, continuous, real-valued functions on S. We shall consider the normed linear spaces $(B(S), \|\cdot\|_\infty)$ and $(C(S), \|\cdot\|_\infty)$ where, as before, if $f \in B(S)$ or $f \in C(S)$, then

$$\|f\|_\infty = \sup\{|f(s)| : s \in S\}.$$

Recall that if $f, g \in B(S)$ and $\|f - g\|_\infty < \epsilon$, say, then this means that $\sup\{|f(s) - g(s)| : s \in S\} < \epsilon$ and so $|f(s) - g(s)| < \epsilon$ for all values of s in S. Figure 5.1 illustrates this fact – the function g must lie within the shaded strip. Now suppose $\{f_n\}$ is a sequence of functions in $B(S)$. What does it mean to say that $\{f_n\}$ converges to f in $B(S)$? At the most elementary level, this means that $\|f_n - f\|_\infty \to 0$ as $n \to \infty$. Thus given $\epsilon > 0$ there is an N in $\mathbb{N}$ such that $\|f_n - f\|_\infty < \epsilon$ whenever $n \geq N$. This can be broken down further by using the definition of the norm: given $\epsilon > 0$ there is an $N \in \mathbb{N}$ such that $|f_n(s) - f(s)| < \epsilon$

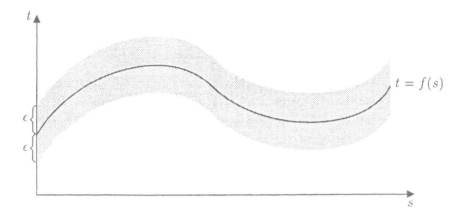

Figure 5.1: Shaded area represents permissible location of graph of g where $\|f - g\| < \epsilon$.

for all $s \in S$, whenever $n \geq N$. So, for a sequence f_n to converge to a function f in $B(S)$, we must require that for each $\epsilon > 0$, there is some 'barrier' N, dependent on ϵ, beyond which each f_n is within ϵ of f, *uniformly* across the whole of the set S. A few examples should help elucidate this concept.

Example Let $S = (0,1)$ and define f_n, $n = 0, 1, 2, \ldots$ by $f_n(s) = s^n$. Then $f_n \in B(S)$, $n = 0, 1, \ldots$. If we fix a value s_0 in S, then as $n \to \infty$, $f_n(s_0) \to 0$. However, the sequence $\{f_n\}_{n=0}^{\infty}$ does *not* converge in $B(S)$ to the zero function. If θ denotes to zero function given by $\theta(s) = 0$, for all $s \in S$, then the claim that $f_n \to \theta$ in $B(S)$ would require that $\|f_n - \theta\|_{\infty} \to 0$ as $n \to \infty$. Let us compute $\|f_n - \theta\|_{\infty}$:

$$\|f_n - \theta\|_{\infty} = \|f_n\|_{\infty} = \sup\{|s^n| : s \in (0,1)\} = 1.$$

Hence $\{f_n\}_{n=0}^{\infty}$ does not converge to θ.

Example Let $S = [0,1]$ and define $f_n, 1 \leq n < \infty$, by

$$f_n(s) = \frac{ns}{1 + n^2 s^2}.$$

Then $f_n \in B(S), 1 \leq n < \infty$, since

$$\|f_n\|_{\infty} = \sup\left\{ \left| \frac{ns}{1 + n^2 s^2} \right| : s \in S \right\} < \sup\{|ns| : s \in S\} \leq n.$$

Again fix s_0 in S. If $s_0 = 0$, then $f_n(s_0) = 0$, $1 \le n < \infty$, while if $s_0 > 0$ then

$$f_n(s_0) = \frac{n s_0}{1 + n^2 s_0^2} < \frac{n s_0}{n^2 s_0^2} = \frac{1}{n s_0}.$$

Hence, for each value of s_0 in $[0,1]$ we have $f_n(s_0) \to 0$. However, it is again the case that the sequence $\{f_n\}_{n=0}^\infty$ does *not* converge in $B(S)$ to the zero function. As in the previous example, we compute $\|f_n - \theta\|_\infty$ where θ is the zero function in $B(S)$:

$$\|f_n - \theta\|_\infty = \|f_n\|_\infty = \sup\{|f_n(s)| : s \in [0,1]\} \ge \left| f_n\left(\frac{1}{n}\right) \right| = 1/2.$$

Hence $\{f_n\}_{n=1}^\infty$ does not converge to θ.

Clearly there is some sort of convergence taking place in each of these examples, but it is not the convergence associated with the normed linear space $(B(S), \|\cdot\|_\infty)$. It is called *pointwise convergence*.

Definition 5.1 *A sequence $\{f_n\}$ in $B(S)$ converges pointwise to f in $B(S)$ if for each s in S, $f_n(s) \to f(s)$.*

Example Let $S = [0,1]$ and define f_n, $1 \le n < \infty$, by $f_n(s) = s/n$. Then

$$\|f_n\|_\infty = \sup\{|s/n| : s \in [0,1]\} = 1/n,$$

and so $\|f_n\|_\infty \to 0$. This means that $\|f_n - \theta\|_\infty \to 0$ as $n \to \infty$, where θ is the zero element in $B(S)$. Hence $\{f_n\}_{n=1}^\infty$ converges to θ in $B(S)$.

This third example shows a sequence $\{f_n\}$ which converges in the sense of convergence in $(B(S), \|\cdot\|_\infty)$. It is also easy to check that $\{f_n\}$ converges pointwise to θ. Convergence of sequences of functions in the normed linear space $(B(S), \|\cdot\|_\infty)$ is called *uniform convergence*.

Definition 5.2 *A sequence $\{f_n\}$ in $B(S)$ converges uniformly to f in $B(S)$ if given $\epsilon > 0$ there exists an $n_0 \in \mathbb{N}$ such that $|f_n(s) - f(s)| < \epsilon$ for all s in S, whenever $n \ge n_0$.*

The subtle difference between pointwise convergence and uniform convergence cannot be emphasized too strongly. It should be clear from the examples and definitions that uniform convergence is a more stringent requirement than pointwise convergence. That is, if the sequence $\{f_n\}$ converges uniformly to f, then it converges pointwise to

f. However, the convergence of $\{f_n\}$ to f pointwise is no guarantee at all of uniform convergence.

If we want to check the two types of convergence then we proceed in essentially different ways. If we suspect that $\{f_n\}$ converges uniformly to f, then we usually evaluate or estimate $\|f - f_n\|_\infty$ and try to show that $\|f - f_n\|_\infty \to 0$. If this is the case, then the sequence $\{f_n\}$ is uniformly and pointwise convergent. If not, we may check for pointwise convergence of $\{f_n\}$ to f. To do this we try to show $f_n(s) \to f(s)$, for each s in S.

Lemma 5.3 *Let S be a subset of $\mathbb{R}$, and let $\{f_n\}$ be a sequence of functions in $B(S)$ which converges uniformly to f in $B(S)$. If each f_n is continuous, then so is f.*

Proof. Take any $s_0 \in S$ and $\epsilon > 0$. Since $\{f_n\}$ is uniformly convergent to f we can find a value $n \in \mathbb{N}$ such that $|f(s) - f_n(s)| < \epsilon/3$ for all s in S. Since f_n is continuous, there is a $\delta > 0$ such that $|f_n(s) - f_n(s_0)| < \epsilon/3$ whenever $|s - s_0| < \delta$. Then for such values of s

$$|f(s) - f(s_0)| \le |f(s) - f_n(s)| + |f_n(s) - f_n(s_0)| + |f_n(s_0) - f(s_0)| \le \epsilon.$$

Hence f is continuous. ∎

This very nice result can be paraphrased by saying that in $B(S)$ a uniformly convergent sequence of continuous functions has a limit which is itself a continuous function. Pointwise convergence is not sufficient to guarantee such a result as the following example shows.

Example Let $S = [0, 1]$ and define f_n, $1 \le n < \infty$, by $f_n(s) = s^n$. Then $f_n \in B(S)$, $1 \le n < \infty$. Also for $s = 1$, $f_n(s) = 1$ while for $0 \le s < 1$, $f_n(s) \to 0$ as $n \to \infty$. Hence $\{f_n\}$ converges pointwise to the function f in $B(S)$ given by

$$f(s) = \begin{cases} 0 & 0 \le s < 1 \\ 1 & s = 1. \end{cases}$$

Each f_n is a continuous function while f is certainly not, so that pointwise convergence cannot guarantee continuity of the limit function.

We now turn to the question of whether $B(S)$ or $C(S)$ possess the two nice properties covered by the previous chapter – completeness and separability.

Theorem 5.4 *The normed linear space* $(B(S), \|\cdot\|_\infty)$ *is complete.*

Proof. Let $\{f_n\}$ be a Cauchy sequence in $B(S)$. We must show that this $\{f_n\}$ converges (in the '$B(S)$ sense') to a function f in $B(S)$. This is done in two steps. Firstly, we identify a likely candidate for f, and then secondly we show that this f is indeed the right choice. To identify the candidate take $\epsilon > 0$. Since $\{f_n\}$ is a Cauchy sequence we can find $n_0 \in \mathbb{N}$ such that whenever $p, q \geq n_0$, then $\|f_p - f_q\|_\infty \leq \epsilon/2$, or equivalently, $|f_p(s) - f_q(s)| \leq \epsilon/2$ for all $s \in S$. This means that for each fixed value of s in S, $\{f_n(s)\}$ is a Cauchy sequence of real numbers and so has a limit which is a real number. We will denote this number by $f(s)$. From the triangle inequality,

$$|\|f_p\| - \|f_q\|| \leq \|f_p - f_q\|,$$

and so $\{\|f_p\|\}$ is a Cauchy sequence of real numbers. It follows that this same sequence is bounded, and from this we may deduce that f lies in $B(S)$. Hence $\{f_n\}$ converges pointwise to f. We show that in fact the convergence is uniform. Fix $p \geq n_0$, and $s \in S$. Choose $q \geq n_0$ so that $|f_p(s) - f_q(s)| \leq \epsilon/2$. Since $f_n(s) \to f(s)$ as $n \to \infty$ we may further constrain q to be sufficiently large to ensure that $|f_q(s) - f(s)| < \epsilon/2$. Then

$$
\begin{aligned}
|f_p(s) - f(s)| &= |f_p(s) - f_q(s)| + |f_q(s) - f(s)| \\
&\leq \frac{\epsilon}{2} + \frac{\epsilon}{2} \\
&= \epsilon.
\end{aligned}
$$

This establishes that for all $s \in S$ and $p \geq n_0$, $|f_p(s) - f(s)| \leq \epsilon$, i.e. for all $p \geq n_0$, $\|f - f_p\|_\infty \leq \epsilon$. Thus $\{f_n\}$ converges to f in $(B(S), \|\cdot\|_\infty)$. ∎

There now follow a series of results illustrating the usefulness of uniform convergence. Some of them concern series of functions rather than sequences. Recall that the infinite series $\sum_{i=1}^\infty f_i$ is always interpreted as the sequence $\{g_n\}_1^\infty$ where $g_n = \sum_{i=1}^n f_i$ – the nth partial sum of the series. Consequently, the statement that the series $\sum_{i=1}^\infty f_i$ converges to f is to be interpreted as the fact that the sequence of partial sums $\{g_n\}$ converges to $f \in B(S)$. Thus if $f_n \in B(S)$, $1 \leq n < \infty$, and the series $\sum_1^\infty f_n$ converges uniformly to f, then this means that given $\epsilon > 0$ there is an $N \in \mathbb{N}$ such that $\|f - \sum_{i=1}^n f_i\|_\infty < \epsilon$ for all $n \geq N$. Alternatively, given $\epsilon > 0$ there is an $N \in \mathbb{N}$ such that for all s in S and for all $n \geq N$, $|f(s) - (f_1(s) + f_2(s) + \ldots f_n(s))| < \epsilon$. The next theorem provides a simple criterion for a series of functions in $B(S)$ to be uniformly convergent.

Theorem 5.5 *(Weierstrass M-test). Let $\{f_n\}_{n=1}^\infty$ be a sequence of functions in $B(S)$ such that the series of real numbers $\sum_{n=1}^\infty \|f_n\|_\infty$ is convergent. Then $\sum_{n=1}^\infty f_n$ converges uniformly on S to a function f in $B(S)$.*

Proof. Let $g_n = \sum_{i=1}^n f_i$, $1 \leq n < \infty$. We have to show that there is an $f \in B(S)$ such that $\|g_n - f\|_\infty \to 0$ as $n \to \infty$. By **5.4**, it will suffice to show that $\{g_n\}$ is Cauchy. Take $\epsilon > 0$. Since $\sum_{n=1}^\infty \|f_n\|_\infty$ is convergent, the sequence $\{r_n\}$ defined by $r_n = \sum_{i=1}^n \|f_i\|_\infty$, where $n = 1, 2, \ldots$, is also convergent. Hence $\{r_n\}$ is a Cauchy sequence. This means that there is an $N \in \mathbb{N}$ such that if $q > p \geq N$ then $|r_q - r_p| \leq \epsilon$, i.e. $\|f_{p+1}\|_\infty + \|f_{p+2}\|_\infty + \ldots + \|f_q\|_\infty \leq \epsilon$. Now for such values of p and q,

$$\|g_q - g_p\|_\infty = \left\| \sum_{i=1}^q f_i - \sum_{i=1}^p f_i \right\|_\infty = \left\| \sum_{i=p+1}^q f_i \right\|_\infty \leq \sum_{i=p+1}^q \|f_i\|_\infty \leq \epsilon.$$

Hence $\{g_n\}$ is a Cauchy sequence as required. ∎

Lemma 5.6 *Suppose the power series $\sum_{n=0}^\infty a_n s^n$ is convergent for some value $s_0 > 0$. Take $0 < r < s_0$. Then the series is uniformly convergent to a continuous function on $[-r, r]$.*

Proof. Since $\sum_{n=0}^\infty a_n s_0^n$ is convergent, $a_n s_0^n \to 0$ as $n \to \infty$. Consequently, there is an $M > 0$ such that $|a_n| s_0^n \leq M$ for all $n \in \mathbb{N}$, and for $n = 0$. Now take $S = [-r, r]$ and define $f_n \in B(S)$ by $f_n(s) = a_n s^n$, $0 \leq n < \infty$. Then

$$\|f_n\|_\infty = \sup_{s \in S} |a_n s^n| = |a_n| r^n = |a_n| s_0^n \left(\frac{r}{s_0}\right)^n \leq M \left(\frac{r}{s_0}\right)^n.$$

Since $r/s_0 < 1$, $\sum_{n=0}^\infty M(r/s_0)^n$ is convergent, and so $\sum_{n=0}^\infty \|f_n\|_\infty$ is convergent. Thus $\sum_{n=0}^\infty f_n$ is a convergent series in $B(S)$ by **5.5**. This is equivalent to the same series being uniformly convergent. That the limit is continuous is a consequence of **5.3**. ∎

Now we give, as nice applications of uniform convergence, two results which depend on knowledge of univariate calculus. Some information on Riemann integration can be found in chapter **11** but you will have to turn to a standard text for information on differentiation (see Spivak [5] for example).

Lemma 5.7 *Let $\{f_n\}$ be a sequence of Riemann-integrable functions on $[a, b]$ and suppose that $\{f_n\}$ converges uniformly to f on $[a, b]$. Then f is Riemann-integrable on $[a, b]$ and*

$$\int_a^b f = \lim_{n \to \infty} \int_a^b f_n.$$

Proof. Take $\epsilon > 0$. Since $\{f_n\}$ converges uniformly to f on $[a, b]$, there is an $N \in \mathbb{N}$ such that for all $n \geq N$ and for all $a \leq s \leq b$ we have $|f(s) - f_n(s)| \leq \epsilon$, that is,

$$f_n(s) - \epsilon \leq f(s) \leq f_n(s) + \epsilon.$$

Using $\overline{\int_a^b} f$ and $\underline{\int_a^b} f$ to represent the upper and lower integrals of f respectively, we see that

$$\overline{\int_a^b} f \leq \overline{\int_a^b} f_n + \epsilon(b - a) = \int_a^b f_n + \epsilon(b - a)$$

and

$$\underline{\int_a^b} f \geq \underline{\int_a^b} f_n - \epsilon(b - a) = \int_a^b f_n - \epsilon(b - a).$$

This gives

$$\overline{\int_a^b} f - \underline{\int_a^b} f \leq 2\epsilon(b - a),$$

and since this inequality is true for all values of $\epsilon > 0$, we must have

$$\overline{\int_a^b} f = \underline{\int_a^b} f.$$

Thus f is Riemann-integrable. The inequalities above may now be rewritten as

$$\int_a^b f_n - \epsilon(b - a) \leq \int_a^b f \leq \int_a^b f_n + \epsilon(b - a)$$

for all $n \geq N$, i.e.

$$\left| \int_a^b f - \int_a^b f_n \right| \leq \epsilon(b - a)$$

for all $n \geq N$. This shows that $\lim_{n \to \infty} \int_a^b f_n = \int_a^b f$. ∎

Lemma 5.8 *Let $S = [a, b]$ and suppose $\{f_n\}$ is a sequence of functions each of which has a continuous derivative on S. Suppose that $\{f_n\}$ converges pointwise to f on S and $\{f_n'\}$ converges uniformly to g on S. Then f is differentiable on (a, b) and $f'(s) = g(s)$ for $a < s < b$.*

Proof. Firstly, since $\{f_n'\}$ converges uniformly to g on S, **5.3** shows that g is continuous on S. Then by **5.7**,

$$\int_a^s g(t)dt = \lim_{n\to\infty} \int_a^s f_n'(t)dt, \quad \text{where} \quad a \leq s \leq b.$$

By a fundamental theorem of calculus,

$$\lim_{n\to\infty} \int_a^s f_n'(t)dt = \lim_{n\to\infty} [f_n(t) - f_n(a)] = f(t) - f(a),$$

and so $\int_a^s g(t)dt = f(s) - f(a)$. Now using a second important theorem of calculus we can conclude that

$$f'(s) = g(s) \quad \text{for} \quad a < s < b. \quad \blacksquare$$

This result has a very nice application to power series. To explain this we need the property established in exercise 5 of this chapter.

Lemma 5.9 *If the power series $\sum_{n=0}^{\infty} a_n s^n$ converges to $g(s)$ for $|s| < R$, then g is differentiable and $g'(s) = \sum_{n=1}^{\infty} n a_n s^{n-1}$ for $|s| < R$.*

Proof. Let

$$f_n(s) = a_0 + a_1 s + \ldots + a_n s^n,$$

so that

$$f_n'(s) = a_1 + 2a_2 s + \ldots + n a_n s^{n-1}.$$

As was remarked prior to this lemma, exercise 5 shows that the sum $\sum_{n=1}^{\infty} n a_n s^{n-1}$ is convergent for each s in $(-R, R)$. Suppose we denote its limit by $h(s)$ for $|s| < R$. Now we show that for any s in $(-R, R)$ the derivative $g'(s)$ is actually $h(s)$. This will establish at a stroke that g is differentiable in $(-R, R)$ *and* has the desired derivative. Take $s_0 \in (-R, R)$ and choose r such that $0 \leq |s_0| < r < R$. By **5.6**, $\sum_{n=1}^{\infty} n a_n s^{n-1}$ is uniformly convergent to $h(s)$ in $[-r, r]$. This means that $\{f_n'\}$ converges uniformly to h on $[-r, r]$. Also one of the hypotheses is that $\{f_n\}$ converges pointwise to g on $(-R, R)$, and so certainly does so on the smaller interval $[-r, r]$. These two statements are just what is needed to infer from **5.8** that $g'(s) = h(s)$ for all s in $(-r, r)$, and so $g'(s_0) = h(s_0)$ as required. $\quad \blacksquare$

The concept of uniform convergence has several applications which at first sight seem quite surprising. We consider one of these now. When the concept of continuity of real-valued functions on an interval $[a, b]$ in $\mathbb{R}$ is introduced, most of us probably feel intuitively that we have a good mental picture of this idea. We visualise a curve that has no 'breaks' or 'jumps'. However, it turns out that some very strange functions can be continuous – ones that we can't begin to visualise. We consider one such example now. The first example of this kind was due to Weierstrass, who shocked the mathematical community when he demonstrated just how barbaric continuous functions could be!

Theorem 5.10 *There exists a continuous function $f : \mathbb{R} \to \mathbb{R}$ which is not differentiable at any point in $\mathbb{R}$.*

Proof. We shall exhibit (more or less) such a function. We define a sequence of functions $\{f_n\}$ in $C(\mathbb{R})$ by first of all setting

$$f_n(s) = |s|, \quad -\frac{1}{2.4^n} < s \leq \frac{1}{2.4^n}.$$

The remaining values are defined by the fact that f_n is periodic with period $1/4^n$. This means that for all $s \in \mathbb{R}$,

$$f_n(s + 1/4^n) = f_n(s),$$

so that the value of $f_n(s)$ is completely determined throughout $\mathbb{R}$ by its values on $(-1/(2.4^n), 1/(2.4^n)]$. One of the functions f_n is graphed in 5.2. Each f_n, $1 \leq n < \infty$, is a continuous function satisfying $|f_n(s)| \leq 1/(2.4^n)$ for all s. Hence $\|f_n\| \leq 1/(2.4^n), 1 \leq n < \infty$, and so the series $\sum_{n=1}^{\infty} f_n$ is uniformly convergent by **5.5**. If we denote by f the sum $\sum_{n=1}^{\infty} f_n$, then by **5.3**, f is continuous. Now we claim that f is not differentiable at any point in R. To see this, take $s \in \mathbb{R}$ and $r \in \mathbb{Z}$. Let us write

$$A_r = \left[\frac{4k}{4^r}, \frac{4k+1}{4^r} \right] \cup \left[\frac{4k+2}{4^r}, \frac{4k+3}{4^r} \right].$$

For $n \geq r$, we have that 4^n is a multiple of 4^r and so

$$f_n \left(s \pm \frac{1}{4^r} \right) = f_n(s).$$

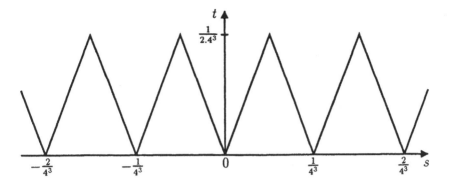

Figure 5.2: The function f_3 in the proof of 5.10.

If $s \in A_r$ then s and $s + \frac{1}{4^r}$ lie under the same line segment of the graph of f_{r-1} (and indeed of f_n, for $n \leq r-1$) so that for $1 \leq n \leq r-1$, the expressions

$$\frac{1}{4^r} \left\{ f_n \left(s + \frac{1}{4^r} \right) - f_n(s) \right\}$$

are either all $+1$ or all -1. Now

$$
\begin{aligned}
\frac{1}{4^r} \left\{ f \left(s + \frac{1}{4^r} \right) - f(s) \right\} &= \sum_{n=1}^{\infty} \frac{1}{4^r} \left\{ f_n \left(s + \frac{1}{4^r} \right) - f_n(s) \right\} \\
&= \sum_{n=1}^{r-1} \frac{1}{4^r} \left\{ f_n \left(s + \frac{1}{4^r} \right) - f_n(s) \right\} \\
&= \pm(r-1).
\end{aligned}
$$

If $s \notin A_r$, then s and $s - \frac{1}{4^r}$ lie under the same line segment of the graph of f_{r-1} and indeed of f_n for $n \leq r-1$, so that for $1 \leq n \leq r-1$, the expressions

$$\frac{1}{4^r} \left\{ f_n \left(s - \frac{1}{4^r} \right) - f_n(s) \right\}$$

are either all $+1$ or all -1. Now again

$$
\begin{aligned}
\frac{1}{4^r} \{ f \left(s - \frac{1}{4^r} \right) - f(s) \} &= \sum_{n=1}^{\infty} \frac{1}{4^r} \left\{ f_n \left(s - \frac{1}{4^r} \right) - f_n(s) \right\} \\
&= \sum_{n=1}^{r-1} \frac{1}{4^r} \left\{ f_n \left(s - \frac{1}{4^r} \right) - f_n(s) \right\} \\
&= \pm(r-1).
\end{aligned}
$$

If $s \in A_r$, then the sequence

$$h_r = \frac{1}{4^r}, \quad r = 1, 2, \dots,$$

is approaching zero, but

$$\frac{f(s + h_r) - f(s)}{h_r} = \frac{1}{4^r}\left\{f\left(s + \frac{1}{4^r}\right) - f(s)\right\} = \pm(r - 1).$$

If $s \notin A_r$ then the sequence

$$h_r = -\frac{1}{4^r}, \quad r = 1, 2, \dots,$$

is approaching zero, but

$$\frac{f(s + h_r) - f(s)}{h_r} = \frac{1}{4^r}\left\{f\left(s - \frac{1}{4^r}\right) - f(s)\right\} = \pm(r - 1).$$

In either case,

$$\lim_{h \to 0} \frac{f(s + h) - f(s)}{h}$$

does not exist, and so f is not differentiable at s. ■

Finally, we spend the remainder of this chapter answering a question about separability. We shall shift our attention to the case $S = [a, b]$, the linear space of all continuous functions on S, denoted by $C(S)$, and the norm $\|f\|_\infty = \sup\{|f(s)| : a \leq s \leq b\}$, $f \in C(S)$. We will show that $C(S)$ is separable by proving a famous result due to Weierstrass, who showed that the polynomials form a dense subset of $C(S)$. This is not an easy result and the reader will need some determination to digest the next few pages. However, it is a very useful result, because it tells us that there is a polynomial p which represents a given continuous function $f \in C(S)$ to a prescribed tolerance $\epsilon > 0$, i.e., $\|f - p\|_\infty < \epsilon$. This is the basis on which most computers and calculators operate. If you press the 'cos' button on a calculator with some given argument, the calculator does not have a chip in it which calculates the cosine of that argument exactly and then truncates to say 8 decimal places so as to fit the number into the limited display. Instead, when the calculator is being designed, the length of the display determines the accuracy to which the cosine function must be obtained. Then the chip evaluates a polynomial which is known to represent cosine to that required accuracy. We begin with a simple observation about $C(S)$.

Theorem 5.11 *The space* $(C(S), \|\cdot\|_\infty)$ *is complete.*

Proof. Let $\{f_n\}$ be a Cauchy sequence in $C(S)$. Then $\{f_n\}$ is Cauchy in $B(S)$ and, since $B(S)$ is complete by **5.4**, $\{f_n\}$ converges to some element f in $B(S)$. Of course, if $\{f_n\}$ converges to f in $B(S)$, then the convergence is uniform convergence, so $\{f_n\}$ converges uniformly to f. Now by **5.3**, $f \in C(S)$ and so $C(S)$ is complete. ∎

Now we set off on something of a journey towards the Weierstrass approximation theorem. It is a journey because the proof is decomposed into several lemmas which look unmotivated in themselves, but which slot together nicely to give the required result. We are going to follow a technique of proof due to S. Bernstein, which is probably the most elegant of the various methods of proof. Until further notice we shall assume $S = [0, 1]$. Given a function f in $C(S)$, we define the nth Bernstein polynomial $B_n f$ by

$$(B_n f)(s) = \sum_{k=0}^{n} \binom{n}{k} f\left(\frac{k}{n}\right) s^k (1 - s)^{n-k}, \quad n = 0, 1, 2, \dots.$$

This is the first time we have encountered such notation, but the Bernstein polynomials depend on the associated function f and the required degree n. We want to indicate both those dependencies. Of course we could write B_n^f or B_f^n or even $_f B_n$. However, there are good reasons for adopting $B_n f$ as the name for the Bernstein polynomial of degree n associated with f, and $(B_n f)(s)$ for its value at the point s.

Lemma 5.12 *Let f_0, f_1 and f_2 be the three functions in $C[0, 1]$ defined by $f_0(s) = 1$, $f_1(s) = s$ and $f_2(s) = s^2$, $0 \le s \le 1$. Then*
(i) $(B_n f_0)(s) = 1$, $0 \le s \le 1$, $n \ge 0$
(ii) $(B_n f_1)(s) = s$, $0 \le s \le 1$, $n \ge 1$
(iii) $(B_n f_2)(s) = \{(n - 1)s^2 + s\}/n$, $0 \le s \le 1$, $n \ge 2$.

Proof. (i) An application of the binomial theorem gives

$$(B_n f_0)(s) = \sum_{k=0}^{n} \binom{n}{k} s^k (1 - s)^{n-k} = (s + (1 - s))^n = 1.$$

(ii) For $n \ge 1$ we have

$$(B_n f_1)(s) = \sum_{k=0}^{n} \binom{n}{k} \left(\frac{k}{n}\right) s^k (1 - s)^{n-k}$$

$$= \sum_{k=1}^{n} \frac{kn!}{nk!(n-k)!} s^k (1-s)^{n-k}$$

$$= \sum_{k=1}^{n} \frac{(n-1)!}{(k-1)!(n-k)!} s^k (1-s)^{n-k}$$

$$= \sum_{k=0}^{n-1} \frac{(n-1)!}{k!(n-k-1)!} s^{k+1} (1-s)^{n-k-1}$$

$$= s \sum_{k=0}^{n-1} \binom{n-1}{k} s^k (1-s)^{n-1-k}$$

$$= s(s+(1-s))^{n-1} = s.$$

Again the binomial theorem was used in the last but one line of this argument.

(iii) For $n \geq 2$,

$$(B_n f_2)(s) = \sum_{k=0}^{n} \binom{n}{k} \left(\frac{k}{n}\right)^2 s^k (1-s)^{n-k}$$

$$= \sum_{k=1}^{n} \frac{k^2 n!}{n^2 k!(n-k)!} s^k (1-s)^{n-k}$$

$$= \sum_{k=1}^{n} \frac{k(n-1)!}{n(k-1)!(n-k)!} s^k (1-s)^{n-k}.$$

Now the astute replacement

$$\frac{k}{n} = \frac{n-1}{n} \frac{k-1}{n-1} + \frac{1}{n},$$

gives

$$(B_n f_2)(s) = \frac{n-1}{n} \sum_{k=2}^{n} \left(\frac{k-1}{n-1}\right) \frac{(n-1)!}{(k-1)!(n-k)!} s^k (1-s)^{n-k}$$

$$+ \frac{1}{n} \sum_{k=1}^{n} \frac{(n-1)!}{(k-1)!(n-k)!} s^k (1-s)^{n-k}.$$

From the argument in (ii) we see that the second term in this expression is s/n and so

$$(B_n f_2)(s) = \frac{n-1}{n} \sum_{k=2}^{n} \frac{(n-2)!}{(k-2)!(n-k)!} s^k (1-s)^{n-k} + \frac{s}{n}$$

$$= \frac{n-1}{n} s^2 \sum_{k=2}^{n} \frac{(n-2)!}{(k-2)!(n-k)!} s^{k-2} (1-s)^{n-k} + \frac{s}{n}$$

$$= \frac{n-1}{n}s^2 \sum_{k=0}^{n-2} \frac{(n-2)!}{k!(n-k-2)!} s^k (1-s)^{n-k-2} + \frac{s}{n}$$

$$= \frac{n-1}{n}s^2 (1 + (1-s))^{n-2} + \frac{s}{n}$$

$$= \frac{n-1}{n}s^2 + \frac{s}{n}. \quad \blacksquare$$

Lemma 5.13 *For $0 \le s \le 1$ and $n \ge 2$, we have*

$$\sum_{k=0}^{n} \binom{n}{k} \left(\frac{k}{n} - s\right)^2 s^k (1-s)^{n-k} = \frac{s(1-s)}{n}.$$

Proof. We have, using **5.12**,

$$\sum_{k=0}^{n} \binom{n}{k} \left(\frac{k}{n} - s\right)^2 s^k (1-s)^{n-k}$$

$$= \sum_{k=0}^{n} \binom{n}{k} \left[\left(\frac{k}{n}\right)^2 - 2\left(\frac{k}{n}\right)s + s^2\right] s^k (1-s)^{n-k}$$

$$= \sum_{k=0}^{n} \binom{n}{k} \frac{k^2}{n^2} s^k (1-s)^{n-k}$$

$$\qquad - 2s \sum_{k=0}^{n} \binom{n}{k} \frac{k}{n} s^k (1-s)^{n-k}$$

$$\qquad + s^2 \sum_{k=0}^{n} \binom{n}{k} s^k (1-s)^{n-k}$$

$$= (B_n f_2)(s) - 2s(B_n f_1)(s) + s^2 (B_n f_0)(s)$$

$$= \left(\frac{n-1}{n}s^2 + \frac{s}{n}\right) - 2s^2 + s^2$$

$$= \frac{s(1-s)}{n}. \quad \blacksquare$$

Lemma 5.14 *For $0 \le s \le 1$ and $\delta > 0$ we have*

$$\sum_{|\frac{k}{n} - s| \ge \delta} \binom{n}{k} s^k (1-s)^{n-k} \le \frac{1}{4n\delta^2},$$

where the notation indicates that the summation is to be taken over all values $k = 0, 1, 2, \ldots, n$ for which $|\frac{k}{n} - s| \ge \delta$.

Proof. We begin by observing that $|\frac{k}{n} - s| \geq \delta$ implies $\frac{1}{\delta^2}(\frac{k}{n} - s)^2 \geq 1$. Then

$$\sum_{|\frac{k}{n}-s|\geq\delta} \binom{n}{k} s^k(1-s)^{n-k} \leq \frac{1}{\delta^2} \sum_{|\frac{k}{n}-s|\geq\delta} \left(\frac{k}{n}-s\right)^2 \binom{n}{k} s^k(1-s)^{n-k}$$

$$\leq \frac{1}{\delta^2} \sum_{k=0}^{n} \left(\frac{k}{n}-s\right)^2 \binom{n}{k} s^k(1-s)^{n-k}$$

$$= \frac{s(1-s)}{\delta^2 n}.$$

Now for $0 \leq s \leq 1$, $s(1-s) \leq \frac{1}{4}$, which gives the desired inequality. ∎

We are now almost ready to prove Weierstrass' theorem. We need to recall some results from univariate real analysis. In fact, these results will also be consequences of the next chapter (see **6.4** and **6.10** for details). Take $f \in C[0,1]$. Then f is bounded on $[0,1]$, that is there is an $M > 0$ such that $|f(s)| < M$ for all $s \in [0,1]$. Secondly, f is uniformly continuous on $[0,1]$. That is, given $\epsilon > 0$ we can find a $\delta > 0$ such that $|f(s) - f(t)| < \epsilon$ whenever $s, t \in [0,1]$ and $|s - t| < \delta$.

Theorem 5.15 *Given $\epsilon > 0$ and $f \in C[0,1]$, there exists an n such that $\|f - B_n f\|_\infty < \epsilon$.*

Proof. Take $\epsilon > 0$ and $f \in C[0,1]$. Choose δ in accordance with the remarks preceeding the theorem, so that $|f(s) - f(t)| < \epsilon/2$ whenever $s, t \in [0,1]$ and $|s - t| < \delta$. Choose M such that $|f(s)| \leq M$ for all $s \in [0,1]$. From **5.12** we can write

$$f(s) = f(s).1 = f(s)(B_n f_0)(s) = f(s)\sum_{k=0}^{n} \binom{n}{k} s^k(1-s)^{n-k}$$

$$= \sum_{k=0}^{n} f(s) \binom{n}{k} s^k(1-s)^{n-k}.$$

Then, for $0 \leq s \leq 1$ and $n \geq 2$,

$$|f(s) - (B_n f)(s)| = \left|\sum_{k=0}^{n} \left[f(s) - f\left(\frac{k}{n}\right)\right] \binom{n}{k} s^k(1-s)^{n-k}\right|$$

$$\leq \sum_{k=0}^{n} \left|f(s) - f\left(\frac{k}{n}\right)\right| \binom{n}{k} s^k(1-s)^{n-k}$$

$$= \sum_{|\frac{k}{n}-s|<\delta} \left| f(s) - f\left(\frac{k}{n}\right) \right| \binom{n}{k} s^k (1-s)^{n-k}$$

$$+ \sum_{|\frac{k}{n}-s|\geq\delta} \left| f(s) - f\left(\frac{k}{n}\right) \right| \binom{n}{k} s^k (1-s)^{n-k}$$

$$\leq \sum_{|\frac{k}{n}-s|<\delta} \frac{\epsilon}{2} \binom{n}{k} s^k (1-s)^{n-k}$$

$$+ \sum_{|\frac{k}{n}-s|\geq\delta} \left\{ |f(s)| + \left| f\left(\frac{k}{n}\right) \right| \right\} \binom{n}{k} s^k (1-s)^{n-k}$$

$$\leq \frac{\epsilon}{2} \sum_{|\frac{k}{n}-s|<\delta} \binom{n}{k} s^k (1-s)^{n-k}$$

$$+ 2M \sum_{|\frac{k}{n}-s|\geq\delta} \binom{n}{k} s^k (1-s)^{n-k}$$

$$\leq \frac{\epsilon}{2} \sum_{k=0}^{n} \binom{n}{k} s^k (1-s)^{n-k}$$

$$+ 2M \sum_{|\frac{k}{n}-s|\geq\delta} \binom{n}{k} s^k (1-s)^{n-k}.$$

Now **5.14** and **5.12** give

$$|f(s) - (B_n f)(s)| \leq \frac{\epsilon}{2} + \frac{M}{2n\delta^2},$$

for all $0 \leq s \leq 1$ and $n \geq 2$. Hence

$$\|f - B_n f\|_\infty \leq \frac{\epsilon}{2} + \frac{M}{2n\delta^2} \quad \text{for } n \geq 2.$$

Now choose n sufficiently large to ensure $M/(2n\delta^2) < \epsilon/2$. Then $\|f - B_n f\|_\infty < \epsilon$. ■

Theorem 5.16 *(Weierstrass). Given $\epsilon > 0$ and $f \in C[a, b]$ there is a polynomial $p \in C[a, b]$ such that $\|f - p\|_\infty < \epsilon$.*

Proof. Suppose $f \in C[a, b]$ and $\epsilon > 0$. Then the function g defined for $0 \leq s \leq 1$ by $g(s) = f(a + (b - a)s)$ lies in $C[0, 1]$. By **5.15**, there is a polynomial $q \in C[0, 1]$ (in fact **5.15** says there is a Bernstein polynomial) such that

$$|g(s) - q(s)| < \epsilon \quad \text{for all} \quad s \in [0, 1].$$

Now define a polynomial $p \in C[a,b]$ by $p(a + (b-a)s) = q(s)$ for $0 \le s \le 1$. Then

$$|f(a + (b-a)s) - p(a + (b-a)s)| < \epsilon \quad \text{for all} \quad s \in [0,1],$$

that is

$$|f(t) - p(t)| < \epsilon \text{ for all } t \in [a,b],$$

so that $\|f - p\|_\infty < \epsilon$. ∎

Although **5.16** is the central result in this area, it is not much more work to show $C[a,b]$ is separable.

Theorem 5.17 *The space $C[a,b]$ is separable.*

Proof. We have from **5.16** that the polynomials form a dense set in $C[a,b]$. We need to exhibit a countable, dense set. Let Q be the set of polynomial functions on $[a,b]$ whose coefficients are rational. Then this set is also dense in $C[a,b]$. To see this take $f \in C[a,b]$ and $\epsilon > 0$. We have to find $q \in Q$ such that $\|f - q\|_\infty < \epsilon$. By **5.16**, choose a polynomial p such that $\|f - p\|_\infty < \epsilon/2$. Write p as $p(s) = \sum_{i=0}^{n} c_i s^i$, $a \le s \le b$. Set $\delta = \max |a|, |b|$. The coefficients $c_i \in \mathbb{R}$ are approximated by numbers $r_i \in \mathbb{Q}$, $0 \le i \le n$, where $|c_i - r_i| \le \epsilon/[2(n+1)\delta^i]$. Set $q(s) = \sum_{i=0}^{n} r_i s^i$, $a \le s \le b$. Then, for $a \le s \le b$,

$$|p(s) - q(s)| = |\sum_{i=0}^{n}(c_i - r_i)s^i| \le \sum_{i=0}^{n}|c_i - r_i|s^i \le \sum_{i=0}^{n}|c_i - r_i|\delta^i$$

$$\le \sum_{i=0}^{n}\frac{\epsilon}{2(n+1)\delta^i}\delta^i = \frac{\epsilon}{2}.$$

Hence $\|p - q\|_\infty \le \epsilon/2$. Then $\|f - q\|_\infty \le \|f - p\|_\infty + \|p - q\|_\infty < \epsilon/2 + \epsilon/2 = \epsilon$. It remains only to check that Q is countable. However, Q is the sum of the sets A_i where $A_i = \{rs^i : a \le s \le b, r \in \mathbb{Q}\}$, $0 \le i < \infty$, and each A_i is countable. It follows (Appendix **A**) that Q is countable. ∎

Exercises

1. Let
$$f_n(s) = \frac{1}{n+s}, \quad s \ge 1.$$
Show that $f_n \to 0$ uniformly on $[0, \infty)$.

2. Let
$$f_n(s) = \frac{ns}{n^2 + s^2}, \quad n \geq 1.$$
Show that $f_n(s) \to 0$ pointwise for all $s > 0$. Does $f_n \to 0$ uniformly (i) on $[0, \infty)$? (ii) on $[0, 1]$?

3. Repeat question 2 for the sequences
$$f_n(s) = \frac{n + s}{n^2 + s}, \quad n \geq 1,$$
and
$$f_n(s) = \frac{s + n}{s^2 + n^2}, \quad n \geq 1.$$

4. Let $f_n(s) = se^{-ns}, n \geq 1$. Show that $f_n \to 0$ uniformly on $[0, \infty)$.

5. Suppose the power series $\sum_{n=0}^{\infty} a_n s^n$ is convergent for $|s| < R$. Show that $\sum_{n=1}^{\infty} n a_n s^{n-1}$ is convergent in this same interval.

6. Let $S = [0, 2\pi]$. Suppose $\sum_{n=1}^{\infty} |a_n|$ is a convergent series of real numbers. Let $f_n \in B(S), 1 \leq n < \infty$, be defined by $f_n(s) = a_n \cos ns$. Show that $\sum_{n=1}^{\infty} f_n$ is uniformly convergent on S.

7. Show that the Taylor series expansion about 0 of the function $f(s) = e^s$ is uniformly convergent in $\mathbb{R}$.

8. This question investigates the Binomial series.

(i) Show that $\sum_{n=0}^{\infty} \binom{\alpha}{n} s^n$ converges for $|s| < 1$. Recall that $\alpha \in \mathbb{R}$ and
$$\binom{\alpha}{n} = \frac{\alpha(\alpha - 1) \dots (\alpha - n + 1)}{n!}.$$

(ii) If $f(s) = \sum_{n=0}^{\infty} \binom{\alpha}{n} s^n$, show that $(1 + s)f'(s) = \alpha f(s)$ where $|s| < 1$.

Show that any function satisfying (ii) must be of the form $f(s) = c(1 + s)^{\alpha}$ for some constant $c \in \mathbb{R}$. Use this fact to establish that
$$(1 + s)^{\alpha} = \sum_{n=0}^{\infty} \binom{\alpha}{n} s^n \quad \text{for } |s| < 1.$$

Here the equality means that the series on the right is uniformly convergent to the function on the left of the equality.

9. Define a sequence of polynomials inductively on $[-1, 1]$ by setting $p_0(t) = 1$ and

$$p_{n+1} = p_n(t) + \frac{1}{2}(t^2 - p_n^2(t)), \quad n = 0, 1, \ldots$$

Show that $0 \leq p_n(t) \leq |t|$ for $-1 \leq t \leq 1$ and hence that $p_n(t) \to |t|$ uniformly on $[-1, 1]$.

10. Let $I = [0, 1]$. Prove that there is no Lipschitz mapping of I onto $I \times I$. [Hint: one can find n^2 points $y_1, \ldots, y_{n^2}$ in $I \times I$ with

$$\|y_i - y_j\| \geq \frac{1}{n-1}, \quad \text{for } i \neq j.$$

Deduce that if f is a Lipschitz mapping from I into $\mathbb{R}^2$ then $f(I)$ has no interior points.]

6

Compactness

We come now to the concept of a compact set. This is rather more subtle than for example the idea of closure. In fact, it is difficult to visualise a compact set without it being indistinguishable from a closed set. We shall encounter the reason for this later in the chapter. The purpose of the notion of a compact set is to distil out the essential quality possessed by a closed, bounded interval in $\mathbb{R}$ which causes every continuous function to attain its bounds. This is such a useful theorem in the analysis of functions of a single real variable, that we should not be surprised to find compactness forming one of the central pillars of modern analysis.

Definition 6.1 *If the set K is contained in some set S, then a family $\mathcal{G}$ of subsets of S is called a covering of K if $K \subseteq \bigcup_{G \in \mathcal{G}} G$. A subfamily of $\mathcal{G}$ (i.e. some of the same elements of $\mathcal{G}$) is called a subcovering if it is itself a covering of K. If S is a normed linear space then the covering is called open (closed) if every member of $\mathcal{G}$ is open (closed) in S.*

Examples

1. The family of sets $\{s\}$ as s ranges over S is a covering of S. No proper subcovering exists.

2. If $S = \mathbb{R}$ then $\{(-n, n); n \in \mathbb{N}\}$ forms an open subcovering.

3. If $S = \mathbb{R}$ then $\{[n, n+1] : n \in \mathbb{Z}\}$ is a closed covering of $\mathbb{R}$. Note for example that $\{(n, n+1] : n \in \mathbb{Z}\}$ is *not* a subcovering since none of the sets $(n, n+1]$ belongs to the original family of

sets. Each set in a subcovering must be in the original list which defines the covering. It is not sufficient for the given set to be a subset of some set in the covering.

Definition 6.2 *Let K be a subset of the normed linear space X. Then K is compact if every open covering of K has a finite subcovering (i.e., one containing only finitely many sets).*

If we are to work only with the definition of compactness and want to prove that a given set K is compact, we must let G be an open covering of K and then show how to select a finite subcovering. When we select G as an open covering it must have no special features, but must be able to play the role of a completely general open covering of K. This is due to the fact that the definition requires us to do something for *all* open coverings of K. Ideally, we would like to pause at this point and examine several examples of compact sets. Unfortunately, this is not the easiest way to proceed. Using the raw definition turns out to be rather hard. What we do is to investigate the relationship between compactness and other concepts in the normed linear space. As we do this, we will begin to build up a picture of some of the properties that go along with compactness.

Despite the pessimism of the previous paragraph, it is easy to see that one simple class of sets is compact. These are the finite sets. Let K be a finite set in the normed linear space X. Suppose $K = \{k_1, k_2, \ldots, k_n\}$. Let G be an open covering of K. Choose G_1 in G such that $k_1 \in G_1$. Choose G_2 in G such that $k_2 \in G_2$. Continuing in this way we choose G_i in G such that $k_i \in G_i, 1 \leq i \leq n$. Then $K \subset \bigcup_{i=1}^{n} G_i$, and so we have constructed the required finite subcovering.

We have already remarked that to establish the compactness of a set K, we must begin with an open covering G which is able to play the role of a completely general open covering. If we already know that K is compact, and wish to deduce some other property, then the situation with regard to the open covering is completely different. If K is compact, then every open covering has a finite subcovering, and we may select very special open coverings, secure in the knowledge that however special our selection may be, there will always be a finite subcovering. A common exploitation of this observation is to cover K with an *increasing* sequence of open sets $G_1 \subseteq G_2 \subseteq G_3 \subseteq \ldots$ so that

$K \subseteq \bigcup_{n=1}^{\infty} G_n$. When we select an open cover, we will have

$$K \subseteq G_{n_1} \cup G_{n_2} \cup \ldots \cup G_{n_j},$$

where $1 \leq n_1 < n_2 < n_3 < \ldots < n_j$. Because the sequence is increasing, we must have $K \subseteq G_{n_j}$. Thus we can conclude in such a situation that there is a number N such that $K \subseteq G_N$.

Lemma 6.3 *If K is a compact subset of the normed linear space X, then K is bounded.*

Proof. Choose any point x_0 in X and set

$$G_n = \{x \in X : \|x - x_0\| < n\}, \quad 1 \leq n < \infty.$$

Then each G_n is open and $G_n \subset G_{n+1}$, $1 \leq n < \infty$. Furthermore, $\bigcup_{n=1}^{\infty} G_n$ is the whole of X. (Recall that $\bigcup_{n=1}^{\infty} G_n = X$ means $\bigcup_{n=1}^{\infty} G_n \subseteq X$ and $X \subseteq \bigcup_{n=1}^{\infty} G_n$. The first of these statements is clear. For the second, take $x \in X$. Then $\|x - x_0\| = r$, say, $r \geq 0$, and $x \in G_n$ whenever $n \geq r$. Hence $X \subseteq \bigcup_{n=1}^{\infty} G_n$.) Now by the remarks preceding the theorem, there is some number N such that $K \subseteq G_N$. This means that

$$\|k - x_0\| < N \quad \text{for all k in K},$$

and so

$$\|k\| \leq \|k - x_0\| + \|x_0\| \leq N + \|x_0\|.$$

Hence K is bounded. ∎

Now we come to the central theorem on compactness.

Theorem 6.4 *Let K be a compact subset of a normed linear space X, and let f be a continuous function from X to $\mathbb{R}$. Then f is bounded and attains its supremum and infimum on K.*

Proof. To show f is bounded on K set

$$G_n = \{x \in X : -n < f(x) < n\}, \quad 1 \leq n < \infty.$$

We can write G_n as $f^{-1}((-n, n))$, and since f is continuous, it follows from **3.7** that G_n is open, $1 \leq n < \infty$. Also, it is again true as with **6.3** that $G_n \subseteq G_{n+1}, 1 \leq n < \infty$ and $\bigcup_{n=1}^{\infty} G_n = X$. Hence there is an

$N \geq 1$ such that $K \subseteq G_N$, which means that $-N < f(k) < N$ for all $k \in K$. Thus f is bounded on K.

We shall now show that f attains its infimum on K. Let $\alpha = \inf\{f(x) : x \in K\}$ and suppose this infimum is *not* attained, that is $f(k) > \alpha$ for all k in K. Set

$$H_n = \left\{ x \in X : f(x) > \alpha + \frac{1}{n} \right\}, \quad 1 \leq n < \infty.$$

Then each H_n can be written as $f^{-1}((\alpha + \frac{1}{n}, \infty))$ and so is open. Also $\{H_n\}_1^\infty$ is an increasing sequence of sets. If k is in K, then $f(k) > \alpha$ by assumption, and so there is an n in N such that $f(k) > \alpha + \frac{1}{n}$. Thus $k \in H_n$ for some $n \in N$, which means that $K \subset \bigcup_{n=1}^\infty H_n$. Now the compactness of K, and the fact that $\{H_n\}$ is increasing imply that $K \subseteq H_N$ for some $N \geq 1$. This in turn means that $f(k) > \alpha + \frac{1}{N}$ for all k in K, which contradicts the fact that α was the infimum of $f(k)$ as k ranges over K. Thus the infimum must be attained.

The demonstration that the supremum is attained can be achieved by making simple modifications to the above argument. However, if we write

$$\sup\{f(k) : k \in K\} = -\inf\{-f(k) : k \in K\},$$

and apply the above argument to the function $-f$ to deduce that the infimum on the right is attained, then it follows immediately that there is a point k_0 in K such that

$$f(k_0) = -\inf\{-f(k) : k \in K\} = \sup\{f(k) : k \in K\}. \quad \blacksquare$$

Corollary 6.5 *Let K be a compact subset of the normed linear space X. If $x \in X$ then there is a point $k_0 \in K$ such that $\|x - k_0\| \leq \|x - k\|$ for all k in K. Alternatively, $\|x - k_0\| = dist(x, K)$, or k_0 is a closest point to x from K.*

Proof. Take x in X and define $f : X \to \mathbb{R}$ by $f(y) = \|x - y\|$, $y \in X$. Then f is continuous and so on the compact set K, f attains its infimum by **6.4**. This means that there is a k_0 in K such that $\|x - k_0\| \leq \|x - k\|$ for all $k \in K$. $\quad \blacksquare$

Corollary 6.6 *Every compact subset in a normed linear space is also closed.*

Proof. Let K be a compact subset of the normed linear space X. Take x in $X \setminus K$. Then by **6.5** there is a k_0 in K such that $\text{dist}(x, K) = \|x - k_0\|$. Since $k_0 \neq x$, $\text{dist}(x, K) > 0$. This means that no point in $X \setminus K$ is a point of closure of K. Hence all points of closure of K must lie in K. That is, K is closed. ∎

We now know from **6.6** and **6.3** that a compact set must be closed and bounded. We shall see soon that this is not enough for a set to be compact – we usually need more than closedness and boundedness. However, we continue to delay our study of concrete examples of compactness. The next two results show how a given set may 'inherit' compactness from another set.

Lemma 6.7 *Every closed subset of a compact set in a normed linear space is compact.*

Proof. Suppose that K is compact in the normed linear space X. Let A be a closed subset of K, and let $\mathcal{G}$ be an open covering of A. Then $\mathcal{G}$ together with $X \setminus A$ is an open covering of K. Therefore, there must be a finite subcovering of K having the form $G_1, G_2, \ldots, G_n, X \setminus A$ where $G_i \in \mathcal{G}, 1 \leq i \leq n$. Since $A \subset K$, this must be an open covering for A as well, but $X \setminus A$ cannot possibly assist us in covering A, so that $G_1, G_2, \ldots, G_n$ must be an open covering for A. Hence A is compact. ∎

Lemma 6.8 *Let f be a continuous mapping from the normed linear space X into a normed linear space Y. If K is a compact subset of X, then $f(K)$ is a compact subset of Y.*

Proof. Let $\mathcal{G}$ be an open covering of $f(K)$. If G is a member of $\mathcal{G}$, then set $H = f^{-1}(G)$. Let $\mathcal{H}$ be the collection of all such sets H as G ranges over $\mathcal{G}$. Then we claim $K \subset \bigcup_{H \in \mathcal{H}} H$. To see this, take $k \in K$. Then $f(k) \in f(K)$ and so $f(k) \in G$ for some G in $\mathcal{G}$, whence it follows that $k \in H$ for some $H \in \mathcal{H}$. Since f is continuous, each H is open in X by **3.7**, and so $\mathcal{H}$ is an open covering of K. Since K is compact, there are sets $H_1, H_2, \ldots, H_n$ in $\mathcal{H}$ which form a finite subcovering of K. Now we claim the corresponding sets $G_1, \ldots, G_n$

where $f^{-1}(G_i) = H_i$ form a finite, open subcovering of $f(K)$. This is established as follows. Take $y \in f(K)$. Then $y = f(k)$ for some k in K. Since $k \in H_j$ for some $1 \leq j \leq n$, $k \in f^{-1}(G_j)$ and so $y \in G_j$ as required. ∎

The following result will turn out to be a small piece of the weaponry needed to attack our main theorem towards the end of this chapter. It is a sort of strengthening of Cantor's Intersection Theorem.

Theorem 6.9 *Let K be a compact subset of a normed linear space X. If $\{F_n\}$ is a decreasing sequence of non-empty, closed subsets of K, then $\bigcap_{n=1}^{\infty} F_n$ is non-empty.*

Proof. Let $\mathcal{G}$ be the family of complements of the sets F_n, $1 \leq n < \infty$. Then each member of $\mathcal{G}$ is open. Suppose that $\bigcap_{n=1}^{\infty} F_n$ is empty. Then $K \subset \bigcup_{G \in \mathcal{G}} G$ and so, since K is compact, a finite number of members of $\mathcal{G}$ cover K, say $G_1, G_2, \ldots G_m$. Thus $K \subset \bigcup_{n=1}^{m} G_n$. Then $\bigcap_{n=1}^{m} F_n$ cannot contain any point of K, and so is empty. This contradicts that fact that $\{F_n\}$ is a decreasing sequence of non-empty sets. ∎

The next result has already been utilised in the previous chapter. (It has to be combined with **6.17** to obtain the assertion with precedes **5.15**.) It tells us that continuous functions are quite often uniformly continuous. This is a nice result, as the prospect of checking the definition of uniform continuity is often quite daunting.

Theorem 6.10 *Let f be a continuous mapping of the normed linear space X into a normed linear space Y. If K is compact subset of X, then f is uniformly continuous on K.*

Proof. Take $\epsilon > 0$. Since f is continuous, then for each k in K there is a $\delta = \delta(k) > 0$ such that whenever $\|x - k\| < \delta(k)$ then $\|f(x) - f(k)\| < \frac{1}{2}\epsilon$. We have emphasised here that continuity can only tell us that δ depends on the location of k in K. What we must show is that the dependence of δ on k is unnecessary – that is, that δ can be chosen independent of which k in K we choose. Set

$$B(k) = \{x \in X : \|x - k\| < \delta(k)/2\}.$$

Then we have $K \subseteq \cup\{B(k) : k \in K\}$. Furthermore, since the $B(k)$ form an open covering of the compact set K, a finite number of these sets cover K. Suppose $B(k_1), \ldots, B(k_n)$ cover K. Set

$$\delta = \min\{\frac{1}{2}\delta(k_i) : 1 \leq i \leq n\}.$$

This value of δ which will perform our desired task. To see this we take $a, b \in K$ with $\|a - b\| < \delta$. Since $b \in B(k_j)$ for some $1 \leq j \leq n$, we have $\|b - k_j\| < \frac{1}{2}\delta(k_j)$. Also

$$\|a - k_j\| \leq \|a - b\| + \|b - k_j\| < \delta + \frac{1}{2}\delta(k_j) \leq \delta(k_j).$$

From the definition of $\delta(k_j)$ we have that

$$\|f(b) - f(k_j)\| < \epsilon/2 \quad \text{and} \quad \|f(a) - f(k_j)\| < \epsilon/2.$$

This gives

$$\|f(b) - f(a)\| \leq \|f(b) - f(k_j)\| + \|f(k_j) - f(a)\| < \frac{1}{2}\epsilon + \frac{1}{2}\epsilon = \epsilon. \quad \blacksquare$$

There is another idea which is closely related to compactness, and perhaps is intuitively more natural than that of open coverings. Recall that a diameter of a set A is $\sup\{\|a_1 - a_2\| : a_1, a_2 \in A\}$.

Definition 6.11 *A subset A of a normed linear space X is totally bounded (or precompact) if for each $\epsilon > 0$, there is a covering of A by finitely many sets of diameter at most ϵ.*

It is left as an exercise for the reader to establish that a totally bounded set is always bounded, so that the concept of a set being totally bounded is *more* demanding than that of boundedness.

Lemma 6.12 *Let A be a subset of a normed linear space X. The following statements are equivalent:*
(i) A is totally bounded
(ii) given $\epsilon > 0$, there exists points $a_1, \ldots, a_n \in A$ such that the closed balls $U_\epsilon(a_i) = \overline{B_\epsilon(a_i)}$ cover A.

Proof. Firstly, suppose A is totally bounded. Then there exist sets $C_1, \ldots, C_n$ with diameters at most ϵ, such that $A \subset \bigcup_{i=1}^n C_i$. Choose $a_i \in C_i \cap A$ whenever this set is non-empty. Then for such values of i, $C_i \subset U_\epsilon(a_i)$ and A is covered by the totality of such sets.

For the reverse implication, take $\epsilon > 0$. Then by (ii) A can be covered by finitely many closed balls of radius $\epsilon/2$. Since these sets have diameter ϵ, A is totally bounded. $\quad \blacksquare$

There now follows another result which makes it easier to visualise a set which is not totally bounded.

Lemma 6.13 *For a set A in a normed linear space X the following are equivalent:*
(i) A is not totally bounded
(ii) for some $\epsilon > 0$, there exists an infinite sequence $\{a_n\}_1^\infty$ in A with $\|a_i - a_j\| \geq \epsilon$ whenever $i, j \geq 1$ and $i \neq j$.

Proof. (i) $\Rightarrow$ (ii) Suppose A is not totally bounded. Then there is an $\epsilon > 0$ such that for every finite subset F of the form $\{a_1, \ldots, a_n\}$ with $a_i \in A$, the balls $U_\epsilon(a_i)$ do not cover A. (Here **6.12**(ii) has been negated.) We now describe how to construct the desired sequence. Choose a_1 in A. Then there is an a_2 in A such that $\|a_1 - a_2\| > \epsilon$, otherwise $U_\epsilon(a_1)$ covers A. Suppose we have constructed $a_1, a_2, \ldots, a_n$ with the desired property $\|a_i - a_j\| \geq \epsilon$, $i \neq j$, $1 \leq i, j \leq n$. Then we must be able to find a_{n+1} in A such that $\|a_i - a_{n+1}\| \geq \epsilon$, $1 \leq i \leq n$, otherwise the closed balls $U_\epsilon(a_i)$, $1 \leq i \leq n$ cover A. This technique defines $\{a_n\}$ inductively.

(ii) $\Rightarrow$ (i) We shall suppose A is totally bounded, and show that the negation of (ii) holds. Take $\epsilon > 0$ and any infinite sequence $\{a_n\}$ in A. Since A can be covered by finitely many sets of diameter at most ϵ, one of these sets must contain two points, a_i and a_j, say. Hence $\|a_i - a_j\| \leq \epsilon$. Since ϵ was arbitrary, the sequence $\{a_n\}$ cannot satisfy (ii). ∎

Now we are very close to our desired objective: demonstrating that there is a rich supply of compact sets in $\mathbb{R}^n$. We shall show first that there are many totally bounded sets in $\mathbb{R}^n$. Then we shall show that totally bounded sets are often quite close to being compact.

Lemma 6.14 *Every bounded set in $\mathbb{R}^n$ (with any one of the norms $\|\cdot\|_1, \|\cdot\|_2, \|\cdot\|_\infty$) is totally bounded.*

Proof. Let A be a bounded set in $\mathbb{R}^n$, with the norm $\|\cdot\|_\infty$. Then there is a real number k such that $\|a\|_\infty \leq k$ for all a in A. If $a = (a_1, \ldots, a_n)$ then $\|a\|_\infty \leq k$ means $-k \leq a_i \leq k$, $1 \leq i \leq n$. Take $\epsilon > 0$ and choose p in N such that $2/p < \epsilon$. Consider the finite set F consisting of points $(\frac{r_1}{p}, \frac{r_2}{p}, \ldots, \frac{r_n}{p})$ where each r_i is an integer in the range $-kp < r_i < kp$, $1 \leq i \leq n$. Let $(a_1, \ldots, a_n) \in A$. For each a_i,

$1 \leq i \leq n$ we will have $|a_i - \frac{r_i}{p}| \leq \frac{1}{p}$ for some integer r_j in the above range. Hence there is an $f \in F$ such that $\|a - f\| \leq \frac{1}{p} < \frac{\epsilon}{2}$. Hence the open balls of radius $\epsilon/2$ with centres as the members of F cover A, and so A is totally bounded. ∎

Now we come to the major result of this section. As we do so, it is worthwhile recalling that the ideas of closure and continuity could each be rewritten in terms of sequences. The same is true of compactness.

Definition 6.15 *Let A be a subset of a normed linear space X. Then A is sequentially compact if every sequence in A has a subsequence which converges to a point in A.*

Recall that a subsequence of the sequence $\{x_n\}$ is usually written $\{x_{n_k}\}$ where this notation means that the subsequence has elements $x_{n_1}, x_{n_2}, x_{n_3}, \ldots$ where $n_1 < n_2 < n_3 < \ldots$, i.e. the subsequence is an 'increasing selection' of members of the original sequence. We shall show now that compactness and sequential compactness are identical concepts in a complete normed linear space. This being so, a legitimate question is why we gave sequential compactness its own special name. A natural progression from abstract analysis is the study of general topology. Here the idea of distance is removed from the initial concept of a topological space and instead everything is derived entirely from a knowledge of open sets. In this setting it turns out that sequential notions are inappropriate, and in several famous cases sequential compactness differs from compactness defined by open coverings. Our care at this point is an attempt to emphasize that there are two distinct concepts here, which happen to coincide in the situation we are studying – that of a complete normed linear space.

Theorem 6.16 *Let A be a subset of a normed linear space X. The following are equivalent:*
(i) A is compact
(ii) A is sequentially compact
(iii) A is complete and totally bounded.

Proof. We shall show that (i) implies (ii), (ii) implies (iii) and then (iii) implies (i). The desired equivalence will then follow.
(i) $\Rightarrow$ (ii) Let A be compact. Let $\{a_n\}$ be a sequence in A. We show how to select a subsequence which converges to a point in A.

Let $R_p = \{a_n : n > p\}$. Then $A \supset \overline{R}_1 \supset \overline{R}_2 \supset \ldots$. Hence $\{\overline{R}_p\}$ is a decreasing sequence of closed, non-empty subsets of A. By **6.9**, $\bigcap_{p=1}^{\infty} \overline{R}_p$ is non-empty. Suppose a_0 is a point in this intersection. Define a subsequence $\{a_{n_k}\}$ as follows: let $a_{n_1} = a_1$. Once $a_{n_1}, \ldots, a_{n_{k-1}}$ are defined, take $a_{n_k} \in R_{n_{k-1}}$ with $\|a_{n_k} - a_0\| \leq 1/i$. Such a choice is always possible because $a_0 \in \overline{R}_{n_{k-1}}$. This defines our subsequence inductively, and $a_{n_k} \to a_0$ as $k \to \infty$.

(ii) $\Rightarrow$ (iii) Suppose A is sequentially compact. To see that A must be complete let $\{a_n\}$ be a Cauchy sequence in A. Since A is sequentially compact, $\{a_n\}$ has a convergent subsequence with limit $a \in A$. Then by **4.12**, $\{a_n\}$ converges to a, and so we conclude that A is complete. Now suppose it is possible for A to be sequentially compact but not totally bounded. Then by **6.13**, there is a sequence $\{a_n\}$ in A and an $\epsilon > 0$ such that $\|a_i - a_j\| \geq \epsilon$ whenever $i, j \geq 1$ and $i \neq j$. Now any subsequence of $\{a_n\}$ must share this separation property and so cannot converge (it is not Cauchy). This contradicts the fact that A was assumed to be sequentially compact.

(iii) $\Rightarrow$ (i) We will suppose A is complete and totally bounded but not compact. The ensuing argument will be a little difficult. If A is not compact, then there is an open covering $\mathcal{G}$ which has no finite subcovering. It will be convenient to call a set *intractable* if it cannot be covered by a finite subfamily of $\mathcal{G}$. Thus A is intractable. Now since A is totally bounded, it can be covered by finitely many closed sets with diameter at most $\frac{1}{2}$. At least one of these, A_1 say, must be intractable. Since A_1 is a subset of a totally bounded set, it is itself totally bounded. Cover A_1 with finitely many sets each of diameter at most $\frac{1}{4}$. One of these, A_2 say, must be intractable. Now the continuation should be clear. We construct in this way a sequence $\{A_n\}_{n=1}^{\infty}$, with $A \supset A_1 \supset A_2 \supset A_3 \supset \ldots$ and $\text{diam}(A_n) \leq 2^{-n}$. By Cantor's intersection theorem (**4.15**) there is a point a_0 in $\bigcap_{n=1}^{\infty} A_n$. We must have that $a_0 \in G$ for some G in $\mathcal{G}$. Since G is open it contains $\{x : \|x - a_0\| \leq 2^{-N}\}$ for some $N > 0$. This forces A_N to lie inside G, since if $a \in A_N$ then because a_0 also lies in A_N, $\|a - a_0\| < 2^{-N}$, and so $a \in G$. The containment $A_N \subset G$ is now a contradiction to the intractibility of A_N. ∎

Corollary 6.17 *A subset of $\mathbb{R}^n$ is compact if and only if it is closed and bounded.*

Proof. From **6.16** a subset of $\mathbb{R}^n$ is compact if and only if it is complete and totally bounded. From the remarks following **6.11** and **6.14**, a set is totally bounded in $\mathbb{R}^n$ if and only if it is bounded. From **4.13**, $\mathbb{R}^n$ is complete and so a subset of $\mathbb{R}^n$ is complete if and only if it is closed. This establishes the result. ■

As advertised earlier, **6.17** provides us with many nice results even when applied to plain old $\mathbb{R}$. For example, a closed interval $[a, b]$ is always compact. Let us reflect for a moment on the information this gives us. Firstly, if a function f defined on $[a, b]$ (and having its values in $\mathbb{R}$, or indeed any normed linear space) is continuous for each s in $[a, b]$ then f is uniformly continuous on $[a, b]$, by **6.10**. In addition if f is a continuous, real-valued function on $[a, b]$ then f is bounded on $[a, b]$ and attains its supremum and infimum there by **6.4**.

Exercises

1. Show that a totally bounded set is bounded. Show that the closure of a totally bounded set is totally bounded. Show that every subset of a totally bounded set is again totally bounded.

2. Using only the definition of compactness, show that:
   (i) (a, b) is not a compact subset of $\mathbb{R}$
   (ii) $\{(x, y) : x^2 + y^2 < 1\}$ is not a compact subset of $\mathbb{R}^2$.

3. Let A be a compact subset of $\mathbb{R}^n$. Let A' denote the embedding of the set A into $\mathbb{R}^m$, $m > n$, using the natural embedding $(x_1, \ldots, x_n) \mapsto (x_1, \ldots, x_n, 0, \ldots, 0)$. Prove that A' is a compact subset of $\mathbb{R}^m$.

4. Show that the union of finitely many compact sets is compact.

5. Let A be a compact subset and B be closed subset of the normed linear space X. If A and B are disjoint sets show that
$$\inf\{\|a - b\| : a \in A, \, b \in B\} > 0.$$

6. Are the following possible?
   (i) A continuous mapping of $[0, 1]$ onto $\mathbb{R}$.
   (ii) A uniformly continuous mapping of $(0, 1)$ onto $\mathbb{R}$.

7. Let f be a uniformly continuous mapping from one normed linear space X into another, Y. Show that the image of a totally bounded set in X is a totally bounded set in Y. Is this true if f is only assumed to be continuous?

8. Which linear subspaces of a normed linear space are compact?

9. Let $\{F_n\}$ be a decreasing sequence of closed sets in a normed linear space X. If $\cap_1^\infty F_n$ is contained in an open set $\mathcal{O}$, show that $F_N \subset \mathcal{O}$ for some N.

10. The following construction shows how to prove directly that $[0,1]$ is a compact subset of $\mathbf{R}$. Let $\mathcal{G} = \{G_\alpha\}$ be a collection of open sets in $\mathbf{R}$ which cover $[0,1]$. Then 0 belongs to some G_α. Since this set G_α is open, $[0,\epsilon) \subset G_\alpha$ for some $\epsilon > 0$. Set $x = \sup\{y \in [0,1]: [0,y]$ is contained in the union of a finite number of members of $\mathcal{G}\}$. Show that in fact $x = 1$.

11. Let A be a compact subset of a normed linear space X. Let x be a point in X and λ a real number. Show that the sets $A_1 = \{x + a : a \in A\}$ and $A_2 = \{\lambda a : a \in A\}$ are both compact.

12. Suppose A is a compact subset of the normed linear space X, and that each x in X has a unique closest point in A. Denote by $\phi(x)$ the closest point to x from A. Prove that ϕ is a continuous mapping.

13. If A is compact in X and B is compact in Y, show that $A \times B$ is compact in $(X \times Y, \|\cdot\|_{X \times Y})$ where

$$\|(x,y)\| = \|x\| + \|y\|, \quad x \in X, y \in Y.$$

14. Let $f : X \to Y$ be such that $f^{-1}(\{y\})$ is compact for each $y \in Y$. Let A_n be a decreasing sequence of closed sets in X with $f(A_n) = Y$. Show that $f(\cap_1^\infty A_n) = Y$. [Hint: Choose $y \in Y$ and consider the sets $A_n \cap f^{-1}(\{y\})$.]

15. Let A be compact and B closed in the normed linear space X. Show that
$$A + B = \{a + b : a \in A, b \in B\}$$
is closed.

7

The contraction mapping theorem

This chapter is a mixture of theory and applications. We have now reached the stage of having enough concepts and attendant theory so that we can discuss some interesting consequences of existing results, as well as continuing to advance our knowledge of the subject. We shall concentrate on mappings between two normed linear spaces X and Y, and prove the well-known 'contraction mapping theorem'.

Definition 7.1 *A mapping f from a normed linear space X into itself is called a contraction if there exists $\lambda < 1$ such that $\|f(x) - f(y)\| \leq \lambda\|x - y\|$ for all x, y in X.*

Note that a contraction is necessarily uniformly continuous on X. In fact, an alternative description of such a mapping would be that it is λ-Lipschitz, where $\lambda < 1$.

Theorem 7.2 *(Contraction Mapping Theorem). Let X be a complete normed linear space, and let f be a contraction on X. Then there is a unique point $x_0 \in X$ such that $f(x_0) = x_0$. The point x_0 is often called the 'fixed point' of f.*

Proof. Let $\lambda < 1$ be such that $\|f(x) - f(y)\| \leq \lambda\|x - y\|$ for all $x, y \in X$. If $f(x_0) = x_0$ and $f(y_0) = y_0$, where $x_0, y_0 \in X$, we have

$$\|x_0 - y_0\| = \|f(x_0) - f(y_0)\| \leq \lambda\|x_0 - y_0\|.$$

Since $\lambda < 1$, this is only possible if $\|x_0 - y_0\| = 0$, and so there exists at most one fixed point.

For the existence of a fixed point, begin with a fixed element x in X. Define a sequence $\{x_n\}_1^\infty$ by $x_1 = f(x)$ and $x_{n+1} = f(x_n)$, $n \geq 1$. Let $\|x - x_1\| = \alpha$. Then

$$\|x_2 - x_1\| = \|f(x_1) - f(x)\| \leq \lambda \|x_1 - x\| = \lambda \alpha.$$

Also,

$$\|x_{n+1} - x_n\| = \|f(x_n) - f(x_{n-1})\| \leq \lambda \|x_n - x_{n-1}\|, \quad n \geq 2.$$

Applying this inequality sufficiently often we obtain

$$
\begin{aligned}
\|x_{n+1} - x_n\| &\leq \lambda \|x_n - x_{n-1}\| \\
&\leq \lambda^2 \|x_{n-1} - x_{n-2}\| \\
&\;\;\vdots \\
&\leq \lambda^n \|x_1 - x\| \\
&= \lambda^n \alpha, \quad n \geq 1.
\end{aligned}
$$

Now for $q > p \geq 1$ we have

$$
\begin{aligned}
\|x_q - x_p\| &\leq \|x_q - x_{q-1}\| + \|x_{q-1} - x_{q-2}\| + \ldots + \|x_{p+1} - x_p\| \\
&\leq \alpha \lambda^{q-1} + \alpha \lambda^{q-2} + \ldots + \alpha \lambda^p \\
&\leq \alpha \frac{\lambda^p}{1 - \lambda}.
\end{aligned}
$$

Since $\lambda^p \to 0$ as $p \to \infty$ this shows that $\{x_n\}_1^\infty$ is a Cauchy sequence. Since X is complete, $\{x_n\}_1^\infty$ converges to a point x_0 in X. Finally, given $\epsilon > 0$, we may choose n such that $\|x_0 - x_n\| \leq \epsilon/2$ and $\|x_{n-1} - x_n\| \leq \epsilon/2$, and so

$$
\begin{aligned}
\|x_0 - f(x_0)\| &\leq \|x_0 - x_n\| + \|x_n - f(x_n)\| \\
&= \|x_0 - x_n\| + \|f(x_{n-1}) - f(x_n)\| \\
&\leq \|x_0 - x_n\| + \lambda \|x_{n-1} - x_n\| \\
&\leq \frac{1}{2}\epsilon + \frac{\lambda}{2}\epsilon \\
&< \epsilon.
\end{aligned}
$$

Since this is true for any value of $\epsilon > 0$, we conclude that $x_0 = f(x_0)$. ∎

We now give an application which illustrates why this theorem is so useful. Other examples, which are less complex, are contained in the exercises. A fundamental question in the theory of ordinary differential equations is the existence of solutions. Suppose we consider a real interval $[a, b]$. Let g be a continuous mapping from $[a, b] \times \mathbb{R}$ into $\mathbb{R}$. Let g also be Lipschitz continuous in its second argument. That is, there is a constant $L > 0$ such that

$$|g(s, t_1) - g(s, t_2)| \le L|t_1 - t_2| \quad \text{for } s \in [a, b] \text{ and } t_1, t_2 \in \mathbb{R}.$$

Then we ask whether, given α in $\mathbb{R}$, there exists a real-valued, differentiable function y defined on $[a, b]$ such that $y(a) = \alpha$ and

$$y'(t) = g(t, y(t)) \qquad \text{for all } t \in [a, b]. \tag{$\dagger$}$$

This is usually called an initial value problem, since the initial value of the function (i.e., its value at the left-hand end point $t = a$) is given and we have to discover the behaviour of the function throughout the rest of the interval. We can, of course, consider initial value problems such that the function g does not satisfy a Lipschitz condition in its second argument. However, this condition enables us to contruct a contraction. As an example, consider the initial value problem

$$y'(t) = t + y(t); \quad y(0) = 0; \quad t \in [0, 1].$$

Here the function g has been chosen to be $g(s, t) = s + t$, and $a = 0$, $b = 1$, $\alpha = 0$. A solution is given by the function y where

$$y(t) = e^t - t - 1.$$

This can easily be checked by substitution, but two important questions present themselves. Is this solution unique, or are there others which satisfy the initial condition and the differential equation? Also, given any g satisfying the stated conditions, does a solution always exist? We shall show that the answer to both these questions is affirmative, using an important and ingenious proof due to Bielecki (see [2]). We are going to use **7.2**, so that three decisions confront us. We must decide which linear space to work in, which norm to employ, and what should be the contraction.

The differential equation $y'(t) = g(t, y(t))$ tells us that on $[a, b]$ the derivative of the function y is given by some other function of t –

actually the somewhat strange looking function $g(t, y(t))$. Thus one version of the fundamental theorem of calculus tells us that for any $t \in [a, b]$ we have

$$\int_a^t g(s, y(s)) \, ds = \int_a^t y'(t) \, dt = y(t) - y(a) = y(t) - \alpha,$$

so that y may be written as

$$y(t) = \alpha + \int_a^t g(s, y(s)) \, ds, \quad t \in [a, b]. \tag{$\ddagger$}$$

This equation is often called an integral equation. Another version of the fundamental theorem of calculus tells us that if y satisfies ($\ddagger$) for some $t \in [a, b]$, then $y'(t) = g(t, y(t))$. (Here $y'(t)$ is understood to mean the right-hand or left-hand derivative when $t = a$ or $t = b$ respectively.) This means that (†) has a unique solution if and only if ($\ddagger$) has a unique solution, so we concentrate on ($\ddagger$).

Our contraction has to be closely connected with ($\ddagger$), and the object we want to conclude (using **7.2**) exists and is unique is the function y which solves ($\ddagger$). From **7.2** this function has to be in the range of our contraction, and so at the very least, the contraction must act on functions. It will also undoubtedly involve integration of such functions, because of the term $\int_a^t g(s, y(s)) ds$. Consequently, our choice of linear space will be $C[a, b]$. Now let the contraction be T. As in chapter 5, we shall denote the image of $u \in C[a, b]$ by Tu rather than $T(u)$. This then allows us to use the notation $(Tu)(t)$ for the value of the function Tu at the point $t \in [a, b]$. We take as the definition of $T : C[a, b] \to C[a, b]$

$$(Tu)(t) = \alpha + \int_a^t g(s, u(s)) \, ds, \quad (t \in [a, b], \ u \in C[a, b]).$$

If y is a fixed point of T, then $Ty = y$ and so

$$(Ty)(t) = y(t) = \alpha + \int_a^t g(s, y(s)) \, ds, \quad (t \in [a, b]).$$

Hence y is a solution of ($\ddagger$).

Let us now review the progress we have made. We have a linear space, $C[a, b]$, and a mapping T whose fixed points are solutions of ($\ddagger$). We must now choose a norm on $C[a, b]$ so that $(C[a, b], \| \cdot \|)$ is complete and T is a contraction. A natural choice would be the supremum norm, since by **5.11**, $C[a, b]$ is complete with this norm. However, such a choice will not permit us to demonstrate that T is a contraction. Instead, we must modify the supremum norm slightly. For $u \in C[a, b]$ we define

$$\|u\|_B = \sup\{\exp(-kL(t - a))|u(t)|; t \in [a, b]\},$$

where $k > 0$ is some constant and L is the Lipschitz constant associated with g. The exercises (again an important ingredient in our discussion, and not an optional extra!) ask you to check that this really does define a norm on $C[a, b]$, and that $C[a, b]$ is complete with respect to this norm.

Theorem 7.3 *Let g be a real-valued, continuous mapping on $[a, b] \times \mathbb{R}$. Suppose there is a constant L such that*

$$|g(s, t_1) - g(s, t_2)| \leq L|t_1 - t_2|, \quad (s \in [a, b], \, t_1, t_2 \in \mathbb{R}).$$

Then the integral equation

$$u(t) = \alpha + \int_a^t g(s, u(s)) \, ds, \quad (t \in [a, b])$$

has a unique solution.

Proof. We maintain the notation of the preamble to the theorem, so we will demonstrate that there is a constant $0 < \lambda < 1$ such that

$$\|Tu - Tv\|_B \leq \lambda \|u - v\|_B$$

for all $u, v \in C[a, b]$. Suppose $u, v \in C[a, b]$. Then, for any t in $[a, b]$,

$$\begin{aligned}
|(Tu)(t) - (Tv)(t)| &= \left| \int_a^t g(s, u(s)) \, ds - \int_a^t g(s, v(s)) \, ds \right| \\
&\leq \int_a^t |g(s, u(s)) - g(s, v(s))| \, ds \\
&\leq \int_a^t L|u(s) - v(s)| \, ds.
\end{aligned}$$

Also from the definition of $\| \cdot \|_B$ we have

$$\|u - v\|_B \geq |u(t) - v(t)| \exp(-kL(t - a)), \quad (t \in [a, b]),$$

so that

$$|u(t) - v(t)| \leq \exp(kL(t - a)) \|u - v\|_B.$$

Thus, for all $t \in [a, b]$,

$$
\begin{aligned}
|(Tu)(t) - (Tv)(t)| &\leq L\|u - v\|_B \int_a^t \exp(kL(s - a))\,ds \\
&= \|u - v\|_B \frac{1}{k} \{\exp(kL(t - a)) - 1\} \\
&\leq \|u - v\|_B \frac{1}{k} \exp(kL(t - a)).
\end{aligned}
$$

Rearranging this equation gives

$$
\exp(-kL(t - a))|(Tu)(t) - (Tv)(t)| \leq \frac{1}{k}\|u - v\|_B.
$$

Now taking the supremum of terms on the left-hand side of this inequality over values of $t \in [a, b]$ gives

$$
\|Tu - Tv\|_B \leq \frac{1}{k}\|u - v\|_B.
$$

Choose the constant k so that $k > 1$. Then T is a contraction and its unique fixed point y is the solution of the integral equation ($\ddagger$). ∎

From the discussion which preceded the theorem, we obtain our desired result about initial value problems, which we state as a corollary.

Corollary 7.4 *Let g be a real-valued, continuous mapping on $[a, b] \times \mathbb{R}$. Suppose there is a constant L such that*

$$
|g(s, t_1) - g(s, t_2)| \leq L|t_1 - t_2|, \quad (s \in [a, b], \ t_1, t_2 \in \mathbb{R}).
$$

Then the differential equation

$$
y'(t) = g(t, y(t)), \quad (t \in [a, b])
$$

with initial condition $y(a) = \alpha$ has a unique solution.

Our whole approach to the discussion and resolution of the existence question for initial value problems is very instructive. In particular, we see that a very special norm was manufactured – one which was 'tailor-made' for the job in hand. We denoted the norm by $\| \cdot \|_B$ in honour of Bielecki, who orginally gave this version of the proof (there are many). We could describe $(C[a, b], \| \cdot \|_B)$ as a 'disposable' normed linear space – one which we use once or twice, then, as it were, throw it away. It is not a space which has a large amount of theory associated with it – indeed one of the exercises shows that it is not sufficiently different from $C[a, b]$ to merit detailed study. However, it did serve our purposes extremely well.

A further point is that it is a worthwhile exercise to examine this same proof when the arguments are only those of classical analysis, and not those of the contraction mapping theorem. A glance at a book like Burkhill [3], will show that many of the messy details have been avoided by our willingness to work with such abstract concepts as complete normed linear spaces and contractions. This in itself should be sufficient to convince the reader that abstract analysis represents a worthwhile investment of time.

In the following exercises, the concept of a mapping being a contraction only on a subset A of the normed linear space X is used. The mapping $f : A \rightarrow X$ is a contraction on A if there exists $0 < \lambda < 1$ such that $\|f(a) - f(b)\| \leq \lambda \|a - b\|$ for all $a, b \in A$.

Exercises

1. In **7.2** the result will no longer hold if we simply assume that $\|f(x) - f(y)\| < \|x - y\|$ for $x, y \in X$. Show that this is true by considering $X = \mathbb{R}$ and $f : X \rightarrow X$ given by

   $$f(s) = \begin{cases} s + \frac{1}{s} & s \geq 1 \\ s - \frac{1}{s-2} & s < 1. \end{cases}$$

2. Calculate constants $a > 0$ and $A > 0$ such that

   $$a\|x\|_\infty \leq \|x\|_B \leq A\|x\|_\infty$$

   for all $x \in C[a, b]$. Deduce (see chapter 4) that $(C[a, b], \|\cdot\|_B)$ is complete.

3. It is possible to prove a limited version of **7.4** using $\|\cdot\|_\infty$ rather than $\|\cdot\|_B$. With the same conditions **7.4** on g, show that there is a value $\beta > a$ such that the differential equation

   $$y'(t) = g(t, y(t)), \quad (t \in [a, b])$$

   with initial value $y(a) = \alpha$ has a unique solution in $[a, \beta]$. (Hint: follow **7.4** as a guide.)

4. Let g be a real-valued, differentiable function on the interval $[a - b, a + b]$. Suppose there is a $\lambda > 0$ such that
   (i) $|g'(t)| \le \lambda < 1$ for all $t \in [a - b, a + b]$
   (ii) $|a - g(a)| \le (1 - \lambda)b$.
   Show that g maps $[a - b, a + b]$ into itself.

5. With the same assumptions as the previous question, show that g is a contraction on $[a - b, a + b]$.

6. Modify the proof of **7.4** to show that if A is complete in X, and f is a contraction mapping A into A, then there is a unique point a in A such that $f(a) = a$.

7. Show that for the setting of question 4, there is a unique point $t_0 \in [a - b, a + b]$ such that $g(t_0) = t_0$. By examining your proof of 6, show that if t is any point in $[a - b, a + b]$ and the sequence $\{t_n\}$ is defined by $t_1 = t$, $t_{n+1} = g(t_n)$, $n \ge 1$, then $t_n \to t_0$. This process is called functional iteration.

8. Give examples of functions which do not have fixed points but do have the following characteristics:
   (i) $f : [0, 1] \to [0, 1]$
   (ii) $f : (0, 1) \to (0, 1)$ and is continuous
   (iii) $f : A \to A$ and is continuous, with $A = [0, 1] \cup [2, 3]$
   (iv) $f : \mathbb{R} \to \mathbb{R}$ and is continuous.

9. Is the following statement true for $F : \mathbb{R} \to [a, b]$? If F is a contraction on $[a, b]$, then F has a unique fixed point which can be obtained by the functional iteration $x_{n+1} = F(x_n)$, $n = 1, 2, \ldots$ starting with any real number x_1.

10. Use the ideas of this chapter to define and evaluate the expression

$$\sqrt{2 + \sqrt{2 + \sqrt{2 + \sqrt{\ldots}}}}$$

[Hint: use iteration!]

11. Let $F : \mathbb{R} \to \mathbb{R}$ be defined by $F(x) = 10 - 2x$. Prove that F has a fixed point. Let x_0 be arbitrary, and define $x_{n+1} = F(x_n)$ for $n = 1, 2, \ldots$. Find a non-recursive formula for x_n. Prove that the method of functional iteration does not produce a convergent

sequence unless x_0 is given a particular value. Determine this value. Why does this example not contradict the contraction mapping theorem?

12. Let $g : \mathbb{R}^2 \to \mathbb{R}^2$ have the components $a : \mathbb{R}^2 \to \mathbb{R}$ and $b : \mathbb{R}^2 \to \mathbb{R}$ where

$$a(s,t) = 1 + \sqrt{t} \quad \text{and} \quad b(s,t) = (s^2 - t^2 - 4s + 2t + 6)/4.$$

Show that the mapping $T : \mathbb{R}^2 \to \mathbb{R}^2$ defined by $T(s,t) = g(s,t)$ has a fixed point which satisfies

$$t^2 = (s-1)^2 \quad \text{and} \quad s^2 + t^2 = 4.$$

13. Show that $f(x) = 3 - x^2$ is a contraction on $[-1/4, 1/4]$, but has no fixed point in this domain. Why does this not contradict **7.2**?

14. Show that $f(x) = -x/2$ is a contraction on $[-2, -1] \cup [1, 2]$ but has no fixed point there. Why does this not contradict **7.2**?

8
Linear mappings

We cannot possibly do justice to the vast literature on linear mappings in one short chapter. However, we shall cover some of the basic properties of these and include one real gem – the uniform boundedness theorem. We have been slowly introducing an alternative notation for mappings between linear spaces. It is this notation which we shall adopt here. Thus a mapping between two linear spaces X and Y will be denoted by a capital letter T, and its action an element of X will be denoted by Tx and not $T(x)$ as would have been customary in earlier chapters. The mapping T is *linear* if

$$T(x_1 + x_2) = Tx_1 + Tx_2, \quad (x_1, x_2 \in X),$$

and

$$T(\alpha x) = \alpha Tx, \quad (\alpha \in \mathbb{R}, x \in X).$$

Since $T\theta_X = T(2\theta_X) = 2T\theta_X$ we see that $T\theta_X = \theta_Y$. Here we have used the notation θ_X to denote the zero element in X. In fact, we will not normally distinguish between θ_X and θ_Y, but write $T\theta = \theta$. Also, note that our declared intention of not using parentheses can only be maintained when T is acting on a single element in X. The problem of continuity for linear mappings is a lot more straightforward than for general mappings. The first two results explain this.

Theorem 8.1 *Let X and Y be normed linear spaces. A linear mapping T from X into Y is continuous if and only if it is continuous at one point.*

Proof. We use the sequential definition of continuity given in **2.10**. Suppose T is continuous at x in X. Consider the point $x' \in X$. Let

$\{x'_n\}$ be a sequence in X with limit x'. Then the sequence $\{x_n\}$, where $x_n = x'_n - x' + x$, $n \geq 1$, converges to x. Since T is continuous at x, $Tx_n \to Tx$, that is, $Tx'_n - Tx' + Tx \to Tx$. This forces us to conclude that $Tx'_n - Tx' \to 0$, that is $Tx'_n \to Tx'$ as required. The reverse implication needs no proof. ∎

Definition 8.2 *Let T be a linear mapping between two normed linear spaces X and Y. If there exists a constant $K > 0$ such that $\|Tx\| \leq K\|x\|$ for all $x \in X$, then T is said to be a bounded linear mapping.*

Note here that the usual abuse of notation has taken place in the inequality $\|Tx\| \leq K\|x\|$. The norm on the left is that of Y, while that on the right is the norm in X.

Example Let $X = (C[0,1], \|\cdot\|_\infty)$ and let Y be $\mathbb{R}$ with the usual norm. Define a linear mapping T from X to Y by

$$Tx = \int_0^1 x(s)\,ds.$$

Then

$$
\begin{aligned}
\|Tx\|_\infty = |Tx| &= \left| \int_0^1 x(s)\,ds \right| \\
&\leq \int_0^1 |x(s)|\,ds \\
&\leq \int_0^1 \max\{|x(t)| : t \in [0,1]\}\,ds \\
&= \int_0^1 \|x\|_\infty\,ds \\
&= \|x\|_\infty.
\end{aligned}
$$

From this calculation we see that T is bounded, and that a suitable choice for K is $K = 1$, that is $\|Tx\|_\infty \leq \|x\|_\infty$ for all $x \in C[0,1]$. Clearly any number which is at least 1 will also serve for K. For example, a correct assertion would be that $\|Tx\|_\infty \leq 7\|x\|_\infty$ for all $x \in C[0,1]$. Again from the above calculation we know that 7 is a far bigger constant than we actually need. This naturally raises the following question. In this concrete example, what is the least value of the constant K which will ensure $\|Tx\|_\infty \leq K\|x\|_\infty$ for all x in $C[0,1]$? In fact the choice $K = 1$ cannot be improved upon. If we consider the function x_0 defined by $x_0(s) = 1$ for all s in $[0,1]$, then

$$\|Tx_0\|_\infty = \left| \int_0^1 1 \, ds \right| = |1| = 1 = \|x_0\|_\infty.$$

Thus any attempt to replace K by a constant smaller than 1 will not make the inequality $\|Tx\|_\infty \leq K\|x\|_\infty$ valid for the function x_0. This best constant is referred to as the norm of the mapping T. In the present case it can be defined as the minimum of all the values K for which $\|Tx\|_\infty \leq K\|x\|_\infty$ for all x in X. However, we have no right to expect that such a minimum will exist for every linear mapping, so in general we take an infimum rather than a minimum.

Definition 8.3 *Let T be a bounded linear mapping of the normed linear space X into a normed linear space Y. Then the norm of T is defined as*

$$\inf\{K : \|Tx\| \leq K\|x\| \text{ for all } x \text{ in } X\}.$$

This number is written as $\|T\|$.

We can with very little trouble replace **8.3** by a more tractable expression.

Lemma 8.4 *Let X and Y be normed linear spaces and let T be a bounded linear mapping from X into Y. Then*

$$\|T\| = \sup_{x \neq \theta} \frac{\|Tx\|}{\|x\|} = \sup_{\|x\| \leq 1} \|Tx\| = \sup_{\|x\| = 1} \|Tx\|.$$

Proof. We shall show that

$$\|T\| = \sup_{x \neq \theta} \frac{\|Tx\|}{\|x\|}$$

and leave the remaining two equalities as exercises. Set

$$K = \sup_{x \neq \theta} \frac{\|Tx\|}{\|x\|}.$$

Then if $x \neq \theta$, we have $\|Tx\|/\|x\| \leq K$ and so $\|Tx\| \leq K\|x\|$. This inequality is also true for $x = \theta$, since $T\theta = \theta$. Thus $\|Tx\| \leq K\|x\|$ for all x in X. Now a glance at **8.3** shows that $\|T\| \leq K$. In fact, $\|T\| = K$, because if it were possible that $\|T\| = K' < K$ then the following contradiction would arise. Since $\sup_{x \neq \theta} \|Tx\|/\|x\| = K$, there is an element x in X such that $\|Tx\|/\|x\| = K - \epsilon$ where $\epsilon = \frac{1}{2}(K - K')$. This gives $\|Tx\| = (K - \epsilon)\|x\| = (K' + \epsilon)\|x\| > K'\|x\|$, which contradicts the definition of K'. ∎

An important consequence of **8.4** is that if T is a bounded linear mapping, then $\|Tx\| \leq \|T\|\|x\|$ for all x in X. Furthermore, the argument contained in the example which led up to the definition of the norm is typical of the computation of the norm of a bounded linear mapping. We first of all establish an upper bound for $\|T\|$, which we try to make as small as possible. Having done this, we try to show it is the smallest bound by picking an element x (or a sequence $\{x_n\}$) for which this bound is attained (or is a supremum).

The astute reader will already have questioned our choice of language in **8.3**. After all, the term *norm* refers to a special mapping from a normed linear space X to $\mathbb{R}$. So if **8.3** defines a norm, then what is the appropriate linear space? Two simple results will explain this.

Theorem 8.5 *A linear mapping T from a normed linear space X into a normed linear space Y is continuous if and only if it is bounded.*

Proof. Suppose T is bounded. It will suffice to show that T is continuous at $\theta \in X$. Let $x_n \to \theta$. Then since $\|Tx_n\| \leq \|T\|\|x_n\|$, we have $Tx_n \to \theta$. Thus T is continuous at θ, and hence by **8.1** is continuous everywhere. Suppose, conversely, that T is continuous. Then T is continuous at θ, and so there is a $\delta > 0$ such that, if $x \in X$ and $\|x\| \leq \delta$, then $\|Tx\| \leq 1$. Now take $x \neq \theta$ in X. Set $x' = \delta x/\|x\|$. Then $\|x'\| = \delta$ and so $\|Tx'\| \leq 1$. This gives

$$1 \geq \|Tx'\| = \left\|T\left(\frac{\delta x}{\|x\|}\right)\right\| = \left\|\frac{\delta}{\|x\|}Tx\right\| = \frac{\delta}{\|x\|}\|Tx\|,$$

so that $\|Tx\| \leq \|x\|/\delta$. Hence T is bounded. ∎

This makes a very easy test for continuity of linear mappings, since boundedness is quite an easy condition to check.

Now let $\mathcal{B}(X,Y)$ represent the set of all bounded, linear mappings from the normed linear space X to the normed linear space Y. Included in $\mathcal{B}(X,Y)$ is the special mapping whose action on every element x in X is to map it to the zero element in Y. We call this the zero mapping.

Theorem 8.6 *Let X and Y be normed linear spaces. Then $B(X,Y)$ is also a normed linear space under the following definitions of addition and multiplication and multiplication by scalars:*

$$(T + S)x = Tx + Sx \quad (T, S \in \mathcal{B}(X,Y), \ x \in X)$$
$$(\alpha T)x = \alpha Tx \quad (T \in \mathcal{B}(X,Y), \ x \in X, \ \alpha \in \mathbb{R}).$$

The norm is as defined in **8.3.**

Proof. There are two points to check. Firstly, we should ascertain that $\mathcal{B}(X,Y)$ forms a linear space under the given definitions. This means verifying that the conditions given in chapter **1** hold. This is a tedious but elementary task, and we content ourselves with showing that our definition of $\|T\|$ for $T \in \mathcal{B}(X,Y)$ really does have the properties required of a norm. Firstly, $\|T\| \geq 0$ for all $T \in \mathcal{B}(X,Y)$. Furthermore, if $\|T\| = 0$ then $\|Tx\| \leq \|T\|\|x\| = 0$ for all x in X, and so $Tx = \theta$ for all x in X. Thus T is the zero mapping. If $\alpha \in \mathbf{R}$, we have, by **8.4,**

$$\|\alpha T\| = \sup_{\|x\|=1} \|(\alpha T)x\| = \sup_{\|x\|=1} \|\alpha Tx\| = |\alpha| \sup_{\|x\|=1} \|Tx\| = |\alpha|\|T\|.$$

Finally, if $T, S \in \mathcal{B}(X,Y)$, then again by **8.4,**

$$
\begin{aligned}
\|T + S\| &= \sup_{\|x\|=1} \|(T + S)x\| \\
&\leq \sup_{\|x\|=1} (\|Tx\| + \|Sx\|) \\
&\leq \sup_{\|x\|=1} \|Tx\| + \sup_{\|x\|=1} \|Sx\| \\
&= \|T\| + \|S\|. \quad \blacksquare
\end{aligned}
$$

There are many times when the situation under consideration is the set of linear mappings from a normed linear space X into itself. In this case the notation $\mathcal{B}(X, X)$ is often abbreviated to $\mathcal{B}(X)$.

Exercises

1. Let S be a subset of $\mathbf{R}$. Show that the mapping $T : B(S) \to \mathbf{R}$ defined by $(Tf)(s) = f(s_0)$, where s_0 is a fixed point in S, is a continuous mapping.

2. Show that the mapping $T : C[0,1] \to C[0,1]$ defined by

$$(Tf)(s) = \int_0^s f, \quad s \in [0,1], \quad f \in C[0,1],$$

is linear and continuous.

3. Show that the shift operator on $B(\mathbb{N})$, defined by

$$(a_1, a_2, a_3, \ldots) \mapsto (0, a_1, a_2, \ldots)$$

is a linear continuous mapping of $B(\mathbb{N})$ into itself.

4. Show that

$$\|T\| = \sup_{\|x\|\le 1} \|Tx\| \quad \text{and} \quad \|T\| = \sup_{\|x\|=1} \|Tx\|$$

both define $\|T\|$.

5. Let

$$A = \begin{pmatrix} a & b \\ c & d \end{pmatrix}$$

be a mapping of $\mathbb{R}^2$ into itself. Show that if $X = (\mathbb{R}^2, \|\cdot\|_\infty)$ then $A \in \mathcal{B}(X)$ and

$$\|A\| = \max(|a| + |b|, |c| + |d|).$$

What happens if the norms $\|\cdot\|_1$ or $\|\cdot\|_2$ are used?

6. Let A be an $n \times n$ matrix, considered as an element of $\mathcal{B}(\mathbb{R}^n)$. Generalise the results of question 4 for the norms $\|\cdot\|_\infty$ and $\|\cdot\|_1$. Can you also deal with $\|\cdot\|_2$?

7. Define $L \in \mathcal{B}(C[0,1])$ by

$$(Lx)(t) = (1-t)x(0) + tx(1), \quad x \in C[0,1].$$

Using the supremum norm, verify that $L \in \mathcal{B}(C[0,1])$ and that $\|L\| = 1$.

8. Let $T \in \mathcal{B}(X)$. Define T^n by $T^n x = T(T^{n-1}x)$, for $n = 2, 3, \ldots$. Show that $\|T^n\| \le \|T\|^n$.

9. Define $T : C[0,1] \to C[0,1]$ by

$$(Tf)(t) = \int_0^t f(s)ds, \quad t \in [0,1].$$

Prove that

$$|(T^n f)(t)| \leq \frac{t^n}{n!}\|f\|, \quad t \in [0,1].$$

Deduce that $\|T^n\| = 1/n!$ Is $\|T^n\| = \|T\|^n$?

10. Let $T \in \mathcal{B}(X,Y)$ where X and Y are normed linear spaces. If A is a bounded subset of X, and T is continuous, show that $T(A)$ is a bounded subset of Y.

11. Show that if $U_X = \{x \in X : \|x\| \leq 1\}$ and $T(U_X)$ is bounded, then T is continuous. [Hint: Take $x \neq 0$ and consider $x/\|x\|$.]

12. Let Π be the space of all polynomials on $[0,1]$ with the supremum norm. For $p \in \Pi$ let Dp denote the polynomial which is the derivative of p. Show that D defines a linear mapping on Π, and that D is discontinuous.

Lemma 8.7 *Let $\mathbb{R}^n$ have any one of the usual norms, and let Y be a normed linear space. Every linear mapping from $\mathbb{R}^n$ to Y is continuous.*

Proof. Let T be a linear mapping from $\mathbb{R}^n$ to Y and use the maximum norm on $\mathbb{R}^n$. Then each element in $\mathbb{R}^n$ can be written as

$$x = (x_1, x_2, \ldots, x_n) = \sum_{i=1}^n x_i e_i,$$

where e_i is the point in $\mathbb{R}^n$ all of whose coordinates are zero except the ith, which is 1. Also, $\|x\|_\infty = \max_{1 \leq i \leq n} |x_i| \geq |x_j|$ for any $1 \leq j \leq n$. The linearity of T means that

$$Tx = T\left(\sum_{i=1}^n x_i e_i\right) = \sum_{i=1}^n T(x_i e_i) = \sum_{i=1}^n x_i T e_i$$

and so

$$\|Tx\| = \|\sum_{i=1}^{n} x_i Te_i\| \leq \sum_{i=1}^{n} \|x_i Te_i\|$$

$$= \sum_{i=1}^{n} |x_i| \|Te_i\|$$

$$\leq \max |x_i| \sum_{i=1}^{n} \|Te_i\|$$

$$= \|x\| \left(\sum_{i=1}^{n} \|Te_i\|\right).$$

Thus in **8.2** we may take $K = \sum_{i=1}^{n} \|Te_i\|$, showing that T is bounded. Finally, T is continuous by **8.5**. ∎

Lemma 8.8 *Let T be an invertible linear mapping of the normed linear space X onto the normed linear space Y. Then T^{-1} is continuous if and only if there is an $\alpha > 0$ such that $\|Tx\| \geq \alpha\|x\|$ for all x in X.*

Proof. Suppose T^{-1} is continuous. Then by **8.5**, T^{-1} is bounded, and so there exists a constant $M > 0$ such that $\|T^{-1}y\| \leq M\|y\|$ for all $y \in Y$. Now for any x in X,

$$\|x\| = \|T^{-1}(Tx)\| \leq M\|Tx\|,$$

and so $\|Tx\| \geq \|x\|/M$ for all x in X. On the other hand, if there is an $\alpha > 0$ such that $\|Tx\| \geq \alpha\|x\|$, then, since T maps X onto Y, given $y \in Y$ there is an $x \in X$ such that $Tx = y$. Hence, $\|Tx\| \geq \alpha\|x\|$ implies $\|y\| \geq \alpha\|T^{-1}y\|$ so that $\|T^{-1}y\| \leq \|y\|/\alpha$ for all $y \in Y$. Thus T^{-1} is bounded, and so by **8.5**, T^{-1} is continuous. ∎

We now make a short but worthwhile digression. The theory of linear mappings allows us to discover that *all* finite-dimensional normed linear spaces are remarkably similar in analytical terms. Let X be an n-dimensional normed linear space, and suppose $x_1, \ldots, x_n$ is a basis for X. There is a natural way of associating points $(\alpha_1, \ldots, \alpha_n)$ in $\mathbb{R}^n$ with points in X, via the basis. Define the mapping $T : \mathbb{R}^n \to X$ by

$$T(\alpha_1, \ldots, \alpha_n) = \sum_{i=1}^{n} \alpha_i x_i.$$

Then T is a linear mapping. One of the exercises shows that in fact T is a continuous bijection from $\mathbb{R}^n$ to X.

Lemma 8.9 *The mapping T as defined in the previous paragraph has a continuous inverse.*

Proof. Suppose T^{-1} is not continuous. Then, using **2.10**, there is a sequence $\{x_n\}$ in X such that $x_n \to x$, but $T^{-1}x_n \not\to T^{-1}x$. By considering the sequence $\{y_n\}$, where $y_n = x_n - x$ for all $n \in \mathbb{N}$, it follows that $y_n \to 0$ but $T^{-1}y_n \not\to 0$. We have to unravel the meaning of this last statement. In fact, there must exist a subsequence of $\{y_n\}$, which we call $\{a_n\}$ and a number $\mu > 0$, such that $a_n \to 0$, but $\|T^{-1}a_n\|_\infty \geq \mu$. Set

$$b_n = \frac{T^{-1}a_n}{\|T^{-1}a_n\|_\infty}, \quad n = 1, 2, \ldots.$$

Then $\|b_n\|_\infty = 1$ and $b_n = T^{-1}c_n$ where

$$c_n = \frac{a_n}{\|T^{-1}a_n\|_\infty} \to 0, \quad \text{as } n \to \infty.$$

The sequence $\{b_n\}$ lies in the compact set $\{z \in \mathbb{R}^n : \|z\| \leq 1\}$ and so contains a convergent subsequence $\{b_{n_k}\}$ with limit $b = (\beta_1, \ldots, \beta_n)$ in $\mathbb{R}^n$. Also, $\|b\|_\infty = 1$. Now, since T is continuous,

$$Tb = \lim_{k \to \infty}[Tb_{n_k}] = \lim_{k \to \infty}\left[\frac{a_{n_k}}{\|T^{-1}a_{n_k}\|_\infty}\right] = 0.$$

Hence

$$0 = Tb = \sum_{i=1}^{n} \beta_i x_i,$$

and, since the x_i are a basis for X (and must therefore be linearly independent) it follows that $\beta_i = 0$ for $i = 1, 2, \ldots, n$. Thus $b = \theta$. This contradicts the fact that $\|b\| = 1$, and therefore T^{-1} must be continuous. ∎

Theorem 8.10 *Let X be an n-dimensional normed linear space. Then there exists a continuous, linear bijection T from $(\mathbb{R}^n, \|\cdot\|_\infty)$ to X such that T has a continuous inverse and*

$$\|T\|^{-1}\|Tz\| \leq \|z\|_\infty \leq \|T^{-1}\|\|Tz\|$$

for all $z \in (\mathbb{R}^n, \|\cdot\|_\infty)$.

Proof. The operator T which was under discussion in the previous lemma can be used. For that operator, T and T^{-1} are bounded, since both are continuous. Hence $\|Tz\| \leq \|T\|\|z\|_\infty$ for all $z \in \mathbf{R}^n$, and this may be rearranged to give $\|z\|_\infty \geq \|T\|^{-1}\|Tz\|$. Now consider the inequality $\|T^{-1}x\|_\infty \leq \|T^{-1}\|\|x\|_\infty$, for all $x \in X$. Writing $x = Tz$ gives $\|T^{-1}Tz\|_\infty \leq \|T^{-1}\|\|Tz\|$, or alternatively, $\|z\|_\infty \leq \|T^{-1}\|\|Tz\|$ for all $z \in \mathbf{R}^n$. $\blacksquare$

This theorem is very reminiscent of the result on the equivalence of the three norms $\|\cdot\|_1$, $\|\cdot\|_2$ and $\|\cdot\|_\infty$ in $\mathbf{R}^n$ (see **2.14**). Its effect is to reduce many aspects of the study of an n-dimensional linear space X to the study of $\mathbf{R}^n$ with any suitable norm. The exercises help the reader to realise some of the special properties of finite-dimensional spaces which follow from **8.10**. In each case they may be summarised by saying that finite-dimensional normed linear spaces often enjoy the same analytical properties as $\mathbf{R}^n$.

Exercises

1. Show that the mapping T defined prior to **8.9** is a linear, continuous bijection.

2. In the notation of **8.10** show that a sequence $\{x_n\}$ in X converges if and only if $T^{-1}x_n$ converges in $(\mathbf{R}^n, \|\cdot\|_\infty)$.

3. In the notation of **8.10** show that a set A is open (closed) in X if and only if $T^{-1}A$ is open (closed) in $(\mathbf{R}^n, \|\cdot\|_\infty)$.

4. Show that if $\|\cdot\|$ is any norm on $\mathbf{R}^n$ then there exist constants a and b such that
$$a\|x\| \leq \|x\|_\infty \leq b\|x\|, \quad \text{for all } x \in \mathbf{R}^n.$$

5. Show that a finite-dimensional normed linear space is always complete.

6. Show that a subset of a finite-dimensional normed linear space is compact if and only if it is closed and bounded.

The next four results take the subject a little further. Some of them will form a key part of our later discussions of the inverse and implicit function theorems, but each is interesting in its own right, and provides experience of working in the space $\mathcal{B}(X, Y)$.

Lemma 8.11 *Let X and Y be normed linear spaces. Suppose that $\{T_n\}$ is a bounded sequence in $\mathcal{B}(X,Y)$ such that $\lim_{n\to\infty} T_n x$ exists for each x in X. If this limit is denoted by Tx then T is a continuous linear mapping of X into Y.*

Proof. The linearity of T follows since for $x, x' \in X$

$$
\begin{aligned}
T(x + x') &= \lim T_n(x + x') \\
&= \lim T_n x + \lim T_n x' \\
&= Tx + Tx',
\end{aligned}
$$

and

$$
T(\lambda x) = \lim T_n(\lambda x) = \lambda \lim T_n x = \lambda(Tx), \quad (\lambda \in \mathbb{R}, x \in X).
$$

Now let $\sup \|T_n\| = M$. Then

$$
\|Tx\| = \lim \|T_n x\| \le M\|x\|
$$

and so $\|T\| \le M$. Hence T is bounded, and so is continuous by 8.5. ∎

Lemma 8.12 *If Y is a Banach space, and X is a normed linear space, then $\mathcal{B}(X,Y)$ is complete.*

Proof. Let $\{T_n\}$ be a Cauchy sequence in $\mathcal{B}(X,Y)$. For each x in X we have

$$
\|T_p x - T_q x\| \le \|T_p - T_q\|\|x\|
$$

for all $p, q \in \mathbb{N}$, and so $\{T_n x\}$ is a Cauchy sequence in Y. Since Y is complete, this sequence converges and we denote its limit by Tx. By 8.11, the mapping T is in $\mathcal{B}(X,Y)$. We have only to establish that $\|T_n - T\| \to 0$ as $n \to \infty$. Take $\epsilon > 0$. There is an n_0 such that whenever $p, q \ge n_0$, we have $\|T_p - T_q\| < \epsilon$. Hence, for each $x \in X$,

$$
\|T_p x - T_q x\| \le \epsilon\|x\|.
$$

Letting $q \to \infty$ in this relation, we obtain, for $p \ge n_0$,

$$
\|T_p x - Tx\| \le \epsilon\|x\|
$$

for all $x \in X$. This means that $\|T_p - T\| < \epsilon$, and so $\{T_n\}$ converges to T, as required. ∎

There now follow three quite deep results. They each take some while to prove, but are well worth the effort needed. The first two concern invertibility.

Lemma 8.13 *Let X be a Banach space. Suppose $T \in \mathcal{B}(X)$ and $\|T\| < 1$. Then the mapping $I - T$ is invertible and*

$$\|(I - T)^{-1} - (I + T)\| \leq \frac{\|T\|^2}{1 - \|T\|}.$$

Proof. Since $\|T^n\| \leq \|T\|^n$ (cf. exercise 8 on page 97) and $\|T\| < 1$, we claim that the elements

$$S_n = I + T + T^2 + \ldots T^n, \quad (n \in \mathbb{N}),$$

form a Cauchy sequence in $\mathcal{B}(X)$. To see this, suppose $p \geq q \geq 1$. Then

$$
\begin{aligned}
\|S_p - S_q\| &= \|T^{q+1} + \ldots + T^p\| \\
&\leq \|T^{q+1}\| + \|T^{q+2}\| + \ldots + \|T^p\| \\
&\leq \|T\|^{q+1} + \|T\|^{q+2} + \ldots + \|T\|^p \\
&\leq \|T\|^{q+1} + \|T\|^{q+2} + \ldots \\
&= \frac{\|T\|^{q+1}}{1 - \|T\|},
\end{aligned}
$$

since $\|T\| < 1$. Given $\epsilon > 0$ we can choose n_0 sufficiently large so that $\|T\|^{q+1}/(1 - \|T\|) < \epsilon$, for $q > n_0$. Then if $p, q > n_0$, we will obtain $\|S_p - S_q\| < \epsilon$. Since $\mathcal{B}(X)$ is complete by **8.12**, $\{S_n\}$ converges to some element $S \in \mathcal{B}(X)$. Now

$$
\begin{aligned}
S_n(I - T) &= S_n - S_n T \\
&= (I + T + \ldots + T^n) - (T + T^2 + \ldots + T^{n+1}) \\
&= I - T^{n+1} \\
&= (I - T)S_n.
\end{aligned}
$$

Since $\|T\| < 1$, $\{T_n\}$ converges to the zero operator in $\mathcal{B}(X)$, and so taking limits in the above inequality gives

$$S(I - T) = I = (I - T)S.$$

Thus S is the inverse of $I - T$. Furthermore

$$
\begin{aligned}
\|S - I - T\| &= \lim_{n \to \infty} \|S_n - I - T\| \\
&= \lim_{n \to \infty} \|T^2 + T^3 + \ldots + T^n\| \\
&\leq \lim_{n \to \infty} \sum_{i=2}^{n} \|T^i\| \\
&\leq \lim_{n \to \infty} \sum_{i=2}^{n} \|T\|^i \\
&= \sum_{i=2}^{\infty} \|T\|^i \\
&= \frac{\|T\|^2}{1 - \|T\|}. \quad \blacksquare
\end{aligned}
$$

Theorem 8.14 *Let X be a Banach space, and let $\mathcal{G}(X)$ be the subset of $\mathcal{B}(X)$ which consists of all elements which have an inverse in $\mathcal{B}(X)$. Then $\mathcal{G}(X)$ is an open subset of $\mathcal{B}(X)$ and the mapping on $\mathcal{G}(X)$ defined by $T \mapsto T^{-1}$ is continuous.*

Proof. Take $T \in \mathcal{G}(X)$. We claim that the ball $B(T, \delta) \subset \mathcal{G}(X)$ provided $\delta < \|T^{-1}\|^{-1}/2$. To see this, take $S \in B(T, \delta)$. Then we can write $S = T + A$ where $\|A\| < \delta < \|T^{-1}\|^{-1}/2$. Since $T + A = T(I + T^{-1}A)$ and $\|T^{-1}A\| \leq \|T^{-1}\|\|A\| < 1/2$, **8.13** may be applied to conclude that $I + T^{-1}A$ is invertible (i.e., its inverse exists and lies in $\mathcal{B}(X)$). Since T is invertible, the composition $S = T(I + T^{-1}A)$ is invertible with inverse in $\mathcal{B}(X)$ as required.

For the continuity of the mapping $T \mapsto T^{-1}$, we continue with the above argument. Observe that $(T + A)^{-1} = (I + T^{-1}A)^{-1}T^{-1}$, so that

$$(T + A)^{-1} - T^{-1} + T^{-1}AT^{-1} = \{(I + T^{-1}A)^{-1} - I + T^{-1}A\}T^{-1}.$$

This gives, using **8.13** and assuming still that $A \in B(T, \delta)$,

$$
\begin{aligned}
\|(T + A)^{-1} - T^{-1} + T^{-1}AT^{-1}\| & \\
&\leq \|(I + T^{-1}A)^{-1} - I + T^{-1}A\|\|T^{-1}\| \\
&\leq \frac{\|T^{-1}A\|^2}{1 - \|T^{-1}A\|}\|T^{-1}\|.
\end{aligned}
$$

Now some simple calculus arguments come to hand. We have

$$\|(T + A)^{-1} - T^{-1} + T^{-1}AT^{-1}\| \leq \frac{\alpha}{1 - \alpha}\|T^{-1}A\|\|T^{-1}\|$$

where $\alpha = \|T^{-1}A\|$, so that $0 \le \alpha < 1/2$. Hence,

$$
\begin{aligned}
\|(T+A)^{-1} - T^{-1} + T^{-1}AT^{-1}\| &\le \max_{\alpha \in [0,1/2]} \left\{ \frac{\alpha}{1-\alpha} \right\} \|T^{-1}A\|\|T^{-1}\| \\
&\le \frac{\frac{1}{2}\|T^{-1}A\|\|T^{-1}\|}{\frac{1}{2}} \\
&= \|T^{-1}A\|\|T^{-1}\|.
\end{aligned}
$$

Thus, for $A \in B(T,\delta)$,

$$
\begin{aligned}
\|(T+A)^{-1} - T^{-1}\| &\le \|(T+A)^{-1} - T^{-1} + T^{-1}AT^{-1}\| \\
&\qquad + \|T^{-1}AT^{-1}\| \\
&\le \|T^{-1}A\|\|T^{-1}\| + \|T^{-1}AT^{-1}\| \\
&\le 2\|T^{-1}\|^2\|A\| \\
&\le 2\|T^{-1}\|^2\delta.
\end{aligned}
$$

Now given $\epsilon > 0$, choose $\delta_1 > 0$ such that $\delta_1 < \|T^{-1}\|^{-1}/2$ and $2\|T^{-1}\|^2\delta_1 < \epsilon$. Then $\|(T+A)^{-1} - T^{-1}\| < \epsilon$ for all $A \in B(T,\delta_1)$, which establishes the required continuity. ∎

The next theorem uses notation which needs some explanation. We want to talk about an arbitrary collection of operators. To do this we use the notation $\{T_\alpha\}_{\alpha \in A}$. Here A represents an *arbitrary* index set. In particular, there is no assumption that A is countable (otherwise our set of operators could be indexed using $\mathbb{N}$). Thus each element T_α is simply assumed to be a member of a perhaps huge collection of operators. For example, if $\alpha \in \mathbb{R}$ then the shift operator on $\mathbb{R}$ associated with α is defined by

$$
T_\alpha(x) = x + \alpha, \quad x \in \mathbb{R}.
$$

The set of all such shift operators T_α as α ranges over $\mathbb{R}$ is an uncountable collection of operators in $\mathcal{B}(\mathbb{R})$.

Theorem 8.15 *Let X be a Banach space, and Y a normed linear space. Let $\{T_\alpha\}_{\alpha \in A}$ be a family of mappings in $\mathcal{B}(X,Y)$. Suppose that for each x in X the set $\{T_\alpha x : \alpha \in A\}$ is bounded. Then the set $\{\|T_\alpha\|\}_{\alpha \in A}$ is bounded.*

Proof. One of our hypotheses is that for each $x \in X$ there is a constant $k(x) > 0$ such that $\|T_\alpha x\| \leq k(x)$ for all $\alpha \in A$. We will direct our energies towards showing that there must be some point x_0 in X, and some $\epsilon > 0$ such that the set $\{T_\alpha x : \alpha \in A\}$ is uniformly bounded on $U_\epsilon(x_0)$. What this means is that there is a $K > 0$ such that for all $\alpha \in A$, and for all x in X with $\|x - x_0\| \leq \epsilon$, we have $\|T_\alpha x\| \leq K$. Before we do this, let us see why this is sufficient for the purpose in hand. Suppose that the constants $K, \epsilon > 0$ and the point $x_0 \in X$ exist. If $y \neq \theta$ then define

$$z = \frac{\epsilon}{\|y\|} y + x_0.$$

Then

$$\|x_0 - z\| = \left\| \frac{\epsilon}{\|y\|} y \right\| = \epsilon,$$

and so $z \in U_\epsilon(x_0)$. Hence by our assumption, $\|T_\alpha z\| \leq K$. Now

$$K \geq \|T_\alpha z\| = \left\| \frac{\epsilon}{\|y\|} T_\alpha y + T_\alpha x_0 \right\| \geq \frac{\epsilon}{\|y\|} \|T_\alpha y\| - \|T_\alpha x_0\|,$$

which gives

$$\|T_\alpha y\| \leq \frac{1}{\epsilon} (K + \|T_\alpha x_0\|) \|y\|.$$

One of the hypotheses of our theorem is that there is some constant $K' > 0$ such that $\|T_\alpha x_0\| \leq K'$ for all $\alpha \in A$. Hence

$$\|T_\alpha y\| \leq \frac{1}{\epsilon}(K + K')\|y\|,$$

and so $\|T_\alpha\| \leq (K + K')/\epsilon$ for all $\alpha \in A$. That is, $\{\|T_\alpha\|\}_{\alpha \in A}$ is bounded, which is the conclusion we seek.

It remains to show that we can find an $x_0 \in X$ and $\epsilon > 0$ such that $\{T_\alpha x : \alpha \in A\}$ is uniformly bounded on $U_\epsilon(x_0)$. We shall suppose to the contrary that no such ball can be found. Choose x_0 in X and $\epsilon_0 > 0$. Within the ball $U_\epsilon(x_0)$ there must exist a point x_1 such that $\|T_{\alpha_1} x_1\| > 1$ for some index $\alpha_1 \in A$. (Otherwise, the set $\{T_\alpha x : \alpha \in A\}$ would be uniformly bounded by 1 on $U_\epsilon(x_0)$.) Since T_{α_1} is bounded, it is continuous (from **8.5**), and so there is some $\epsilon_1 > 0$ such that $\|T_{\alpha_1} x\| > 1$ for all $\|x - x_1\| < \epsilon_1$. This can be restated by saying that there is an open ball $B_{\epsilon_1}(x_1)$ such that $\|T_{\alpha_1} x\| > 1$ for all

$x \in B_{\epsilon_1}(x_1)$. We will assume $\epsilon_1 < 1$. This argument is now repeated. Since $\{T_\alpha x : \alpha \in A\}$ cannot be uniformly bounded on $B_{\epsilon_1}(x_1)$, there exists a point $x_2 \in B_{\epsilon_1}(x_1)$ such that $\|T_{\alpha_2} x_2\| > 2$ for some index α_2. Since T_{α_2} is continuous, there is an open ball $B_{\epsilon_2}(x_2)$ such that $\|T_{\alpha_2} x\| > 2$ for all $x \in B_{\epsilon_2}(x_2)$. We can suppose that $0 < \epsilon_2 < 1/2$. Continuing in this way, we obtain a sequence of points $x_1, x_2, \ldots$, indices $\alpha_1, \alpha_2, \ldots$ and positive numbers $\epsilon_1, \epsilon_2, \ldots$ such that

$$B_{\epsilon_n}(x_n) \subset B_{\epsilon_{n-1}}(x_{n-1}) \quad n = 2, 3, \ldots,$$

$\|T_{\alpha_n} x_n\| > n$, and $\epsilon_n < 1/n$. Now $\{\overline{B_{\epsilon_n}(x_n)}\}$ is a decreasing sequence of non-empty, closed subsets of the complete normed linear space X. Furthermore, $\mathrm{diam} \overline{B_{\epsilon_n}(x_n)} \to 0$ as $n \to \infty$. Thus by **4.15** there exists a point $z \in \cap_{n=1}^\infty \overline{B_{\epsilon_n}(x_n)}$. Since $z \in \overline{B_{\epsilon_n}(x_n)}$ for all n, this forces $\|T_{\alpha_n} z\| \geq n$, for $n \geq 1$, and thus contradicts the assumption that $\{T_\alpha z : \alpha \in A\}$ is a bounded set. ∎

There are numerous applications of the above result, and we turn now to one concerned with Fourier series. It will be convenient to work in the linear space $C_{2\pi}$. This space consists of all real-valued, continuous functions defined on $\mathbb{R}$, which are also 2π-periodic. That is, if $x \in C_{2\pi}$ then

$$x(t + 2\pi) = x(t) \quad \text{for all } t \in \mathbb{R}.$$

Consequently, knowledge of the behaviour of $x \in C_{2\pi}$ on any interval of length 2π is sufficient for us to be in possession of complete knowledge of x. The functions $\sin$ and $\cos$ are simple examples of 2π-periodic, continuous functions. The supremum norm

$$\|x\|_\infty = \max\{|x(t)| : 0 \leq t \leq 2\pi\}$$

will be used in $C_{2\pi}$. A useful practical tool for approximating functions in $C_{2\pi}$ is the Fourier series. Formally, if $x \in C_{2\pi}$, then this is defined to be the infinite series

$$\frac{1}{2}a_0 + \sum_{k=1}^\infty (a_k \cos kt + b_k \sin kt)$$

where

$$a_k = \frac{1}{\pi} \int_{-\pi}^\pi x(s) \cos ks \, ds, \quad k = 0, 1, 2, \ldots$$

and

$$b_k = \frac{1}{\pi} \int_{-\pi}^{\pi} x(s) \sin ks \, ds, \quad k = 1, 2, \ldots.$$

We describe this definition as 'formal' because we are not making any assertion at the moment about the convergence of the series. Indeed that is one of the major questions about the Fourier series. Given a function $x \in C_{2\pi}$, does its Fourier series converge uniformly? After a long period (almost 50 years), when it was conjectured that the answer to this question was affirmative, Du Bois-Reymond produced in 1859 an example of a continuous function whose Fourier series diverged at a point. We will show how the uniform boundedness theorem can be used to deduce the existence of such a function, although it will fail to supply knowledge of the actual form of the function. Before considering the Fourier series specifically, we point out a simple corollary to **8.15**, which shows how this result provides information about the convergence of linear processes.

Corollary 8.16 *Let X be a Banach space and $\{L_n\}_1^\infty$ a sequence of bounded, linear mappings from X into itself. If $\lim_n \|x - L_n x\| = 0$ for all x in X, then $\{\|L_n\|\}_1^\infty$ is a bounded sequence (of real numbers).*

Proof. If $\lim_n \|x - L_n x\| = 0$ for each $x \in X$, then there must be some constant $k(x)$ such that

$$\|x - L_n x\| \leq k(x) \quad \text{for } n \geq 1.$$

Then for each x in X, we have

$$\|L_n x\| = \|x - x + L_n x\| \leq \|x\| + \|x - L_n x\| \leq \|x\| + k(x), \quad n \geq 1.$$

This means that for each $x \in X$, the set $\{\|L_n x\| : n \in \mathbb{N}\}$ is bounded. It follows from **8.15** that $\{\|L_n\|\}_1^\infty$ is bounded sequence. ∎

The negation of **8.16** will be that if $\{\|L_n\|\}_1^\infty$ is unbounded, then there must be at least one x in X such that $\{L_n x\}$ does not converge to x in X. It is this result which will be employed to show the existence of a function $x \in C_{2\pi}$ for which the Fourier series does not converge. To apply **8.16** we need to construct a sequence of linear mappings. Somehow the 'limit' of this sequence in some sense must be the formal Fourier series for the given function in $C_{2\pi}$. Recall that the assertion that the Fourier series of $x \in C_{2\pi}$ converges uniformly to x means that the *sequence* of partial sums

$$\frac{1}{2}a_0 + \sum_{k=1}^{n}(a_k \cos kt + b_k \sin kt)$$

converges uniformly to x as $n \to \infty$. It is these partial sums which define suitable linear mappings. For $n \geq 1$, define the linear mapping $S_n \in \mathcal{B}(C_{2\pi})$ by

$$(S_n x)(t) = \frac{1}{2}a_0 + \sum_{k=1}^{n}(a_k \cos kt + b_k \sin kt), \quad n \geq 1.$$

Recall that the notation here is chosen so that S_n is the linear mapping, $S_n x$ is the element of $C_{2\pi}$ which is produced by the action of S_n on the element x in $C_{2\pi}$, and $(S_n x)(t)$ is the value of the function $S_n x$ at the point $t \in \mathbf{R}$. Our stated goal is quite some way off, and will involve very beautiful applications and concepts from the theory of normed linear spaces. We will have to verify that $\{\|S_n\|\}$ is an unbounded sequence, and this will involve several arguments which discuss the nature of these linear mappings S_n in some detail. These arguments will become quite messy, and so notational conveniences will help.

Our first move is to stand back form the problem and consider what sort of functions the linear mappings S_n may be allowed to act upon. That amounts to asking 'For what sort of functions are the coefficients a_k and b_k defined?' Since their definition involved integration, the continuity of the function x is sufficient to make the product of x with the cosine and sine functions of multiple angles a (Riemann) integrable function. However, something weaker than the continuity of x will suffice, since there are a lot of (Riemann) integrable functions which are not continuous. If x is a bounded, integrable, 2π-periodic function, then the theory of the Riemann integral tells us that the functions which form the integrands in the expressions for the a_k and b_k are indeed Riemann-integrable, and so in this case these coefficients and S_n will be well-defined. In fact, the set of all bounded, integrable, 2π-periodic functions, which we will denote by $F_{2\pi}$, forms a linear space, although that will not concern us here. (Remember that all we have to check is that the sum of two functions in $F_{2\pi}$ is again a bounded, integrable, 2π-periodic function, and that any multiple of an element of $F_{2\pi}$ is again an element of $F_{2\pi}$. These results are elementary consequences of the definition of the Riemann integral. If you are uncertain about the theory of Riemann integration, then chapter **11** gives a review of this subject.)

Lemma 8.17 *Let x be a bounded, integrable, 2π-periodic function. Then the Fourier series mapping S_n has the following integral form:*

$$(S_n x)(t) = \frac{1}{\pi} \int_{-\pi}^{\pi} x(t+s) \frac{\sin(n+\frac{1}{2})s}{2\sin\frac{1}{2}s} \, ds, \quad t \in \mathbb{R}, \, n \geq 0.$$

Proof. The proof is manipulative in character. For $x \in F_{2\pi}$ we have

$$
\begin{aligned}
(S_n x)(t) &= \frac{1}{2}a_0 + \sum_{k=1}^{n}(a_k \cos kt + b_k \sin kt) \\
&= \frac{1}{2\pi} \int_{-\pi}^{\pi} x(s) \, ds \\
&\quad + \sum_{k=1}^{n} \left[\begin{array}{l} \cos kt \, \frac{1}{\pi} \int_{-\pi}^{\pi} x(s) \cos ks \, ds \\ \qquad + \sin kt \, \frac{1}{\pi} \int_{-\pi}^{\pi} x(s) \sin ks \, ds \end{array} \right] \\
&= \frac{1}{\pi} \int_{-\pi}^{\pi} x(s) \left[\frac{1}{2} + \sum_{k=1}^{n}(\cos kt \cos ks + \sin kt \sin ks) \right] ds.
\end{aligned}
$$

An application of the trigonometric identity

$$\cos(A - B) = \cos A \cos B + \sin A \sin B$$

gives

$$(S_n x)(t) = \frac{1}{\pi} \int_{-\pi}^{\pi} x(s) \left[\frac{1}{2} + \sum_{k=1}^{n} \cos k(s - t) \right] ds.$$

Each function in the integrand is periodic with period 2π, and so integration over *any* interval of length 2π will give the same result. This argument, followed by a simple change of variable explains the next few lines:

$$
\begin{aligned}
(S_n x)(t) &= \frac{1}{\pi} \int_{-\pi}^{\pi} x(s) \left[\frac{1}{2} + \sum_{k=1}^{n} \cos k(s - t) \right] ds \\
&= \frac{1}{\pi} \int_{-\pi+t}^{\pi+t} x(s) \left[\frac{1}{2} + \sum_{k=1}^{n} \cos k(s - t) \right] ds \\
&= \frac{1}{\pi} \int_{-\pi}^{\pi} x(s + t) \left[\frac{1}{2} + \sum_{k=1}^{n} \cos ks \right] ds.
\end{aligned}
$$

If we compare this last integral with the statement of the lemma, then we see we must verify that

$$\frac{1}{2} + \sum_{k=1}^{n} \cos ks = \frac{\sin(n + \frac{1}{2})s}{2 \sin \frac{1}{2}s}, \quad s \in [-\pi, \pi].$$

There is a small difficulty here, in that the expression on the right-hand side is not defined for $s = 0$. However, it is simple to check that

$$\lim_{s \to 0} \frac{\sin(n + \frac{1}{2})s}{2 \sin \frac{1}{2}s} = n + 1/2, \quad n \geq 0,$$

so that our integrand in the lemma, and our equation above are to be understood as involving the continuous function defined by

$$\frac{\sin(n + \frac{1}{2})s}{2 \sin \frac{1}{2}s}, \quad s \neq 0,$$

and $n + 1/2$ when $s = 0$. Thus we must establish that

$$\frac{1}{2} + \sum_{k=1}^{n} \cos ks = \begin{cases} (\sin(n + \frac{1}{2})s)/(2 \sin \frac{1}{2}s) & s \neq 0, \quad n \geq 0 \\ n + 1/2 & s = 0, \quad n \geq 0. \end{cases}$$

This inequality is elementary for $s = 0$ and so for $s \neq 0$ we must establish

$$\sin \frac{1}{2}s + 2 \sin \frac{1}{2}s \sum_{k=1}^{n} \cos ks = \sin \left(n + \frac{1}{2}\right)s, \quad n \geq 0.$$

By using the identity $2 \cos A \sin B = \sin(A + B) - \sin(A - B)$ we have

$$\begin{aligned}
\sin \frac{1}{2}s + 2 \sin \frac{1}{2}s \sum_{k=1}^{n} \cos ks &= \sin \frac{1}{2}s + 2 \sum_{k=1}^{n} \sin \frac{1}{2}s \cos ks \\
&= \sin \frac{1}{2}s \\
&\quad + \sum_{k=1}^{n} \left[\sin \left(k + \frac{1}{2}\right)s - \sin \left(k - \frac{1}{2}\right)s \right] \\
&= \sin \left(n + \frac{1}{2}\right)s. \quad \blacksquare
\end{aligned}$$

We often use the notation

$$(S_n x)(t) = \frac{1}{2\pi} \int_{-\pi}^{\pi} x(t + s) D_n(s)\, ds$$

where D_n is the *Dirichlet kernel* defined by

$$D_n(t) = \frac{\sin(n + \frac{1}{2})t}{\sin \frac{1}{2}t}, \quad t \in \mathbf{R}.$$

Once again, limiting values are assumed at multiples of 2π. It will be convenient to use two different norms on $F_{2\pi}$ for a while. We retain the supremum norm, so that

$$\|x\|_\infty = \sup\{|x(t)| : -\pi \le t \le \pi\}.$$

We also define

$$\|x\|_1 = \frac{1}{2\pi} \int_{-\pi}^{\pi} |x(s)| \, ds.$$

Lemma 8.18 *For any $x \in F_{2\pi}$ and $n \ge 0$,*

$$\|S_n x\|_\infty \le \|x\|_\infty \|D_n\|_1 \le \left(n + \frac{1}{2}\right) \|x\|_\infty.$$

Consequently, S_n is a bounded linear mapping on $(F_{2\pi}, \|\cdot\|_\infty)$.

Proof. Let x be in $F_{2\pi}$. Then, for any $t \in [-\pi, \pi]$ we have

$$
\begin{aligned}
|(S_n x)(t)| &= \left| \frac{1}{2\pi} \int_{-\pi}^{\pi} x(t+s) D_n(s) \, ds \right| \\
&\le \frac{1}{2\pi} \int_{-\pi}^{\pi} |x(t+s)| |D_n(s)| \, ds \\
&\le \frac{1}{2\pi} \int_{-\pi}^{\pi} \|x\|_\infty |D_n(s)| \, ds \\
&= \|x\|_\infty \frac{1}{2\pi} \int_{-\pi}^{\pi} |D_n(s)| \, ds \\
&= \|x\|_\infty \|D_n\|_1.
\end{aligned}
$$

Thus

$$\|S_n x\|_\infty = \max_{t \in [-\pi, \pi]} |(S_n x)(t)| \le \|x\|_\infty \|D_n\|_1.$$

The second inequality in the lemma follows from the fact that for all $t \in [-\pi, \pi]$ we have $|D_n(t)| \le n + 1/2$ so that

$$\|D_n\|_1 = \frac{1}{2\pi} \int_{-\pi}^{\pi} |D_n(s)| \, ds \le \frac{1}{2\pi} \int_{-\pi}^{\pi} \left(n + \frac{1}{2}\right) ds = n + \frac{1}{2}. \quad \blacksquare$$

The above result gives a very crude bound of $n + 1/2$ for $\|S_n\|$. If we work a little harder then $\|S_n\|$ can be computed exactly in terms of the Dirichlet kernel.

Lemma 8.19 *Let $X = (F_{2\pi}, \|\cdot\|_\infty)$. Then the operator S_n in $\mathcal{B}(X)$ has norm $\|S_n\| = \|D_n\|_1$.*

Proof. It follows from **8.18** that $\|S_n\| \leq \|D_n\|_1$. We shall exhibit a function $\hat{x}$ in $F_{2\pi}$ such that $\|\hat{x}\|_\infty = 1$ and $\|S_n\hat{x}\|_\infty \geq \|D_n\|_1$. This will be sufficient to prove that $\|S_n\| = \|D_n\|_1$. The function D_n is only zero in $[-\pi, \pi]$ when its numerator is zero and the denominator is non-zero. Since

$$D_n(t) = \frac{\sin(n + \frac{1}{2})t}{\sin \frac{1}{2}t}, \quad t \in \mathbb{R},$$

we see that $D_n(t) = 0$ if and only if

$$t = \pm\frac{k\pi}{n + 1/2}, \quad k = 1, 2, \ldots, n.$$

Define the function $\hat{x}$ on $[-\pi, \pi]$ by

$$\hat{x}(t) = \begin{cases} 1 & D_n(t) > 0 \\ 0 & D_n(t) = 0 \\ -1 & D_n(t) < 0 \end{cases}.$$

This function is sketched in 8.1 for $n = 3$. The construction makes $\hat{x}$ a bounded integrable function whose sign agrees with that of D_n on $[-\pi, \pi]$. Now extend $\hat{x}$ to the whole of $\mathbb{R}$ by periodicity. Then $\hat{x} \in F_{2\pi}$ and $\|\hat{x}\|_\infty = 1$. By **8.17** we obtain

$$
\begin{aligned}
\|S_n\hat{x}\|_\infty &= \max_{t\in[-\pi,\pi]} |(S_n\hat{x})(t)| \\
&\geq |(S_n\hat{x})(0)| \\
&= \left| \frac{1}{2\pi} \int_{-\pi}^{\pi} \hat{x}(s)D_n(s)\, ds \right| \\
&= \left| \frac{1}{2\pi} \int_{-\pi}^{\pi} |D_n(s)|\, ds \right| \\
&= \|D_n\|_1.
\end{aligned}
$$

Now from **8.18** and **8.4** we have

$$\|S_n\| = \sup_{x\neq\theta} \frac{\|S_n x\|_\infty}{\|x\|_\infty} \leq \|D_n\|_1.$$

But the previous argument shows that

$$\|S_n\| = \sup_{x\neq\theta} \frac{\|S_n x\|_\infty}{\|x\|_\infty} \geq \frac{\|S_n\hat{x}\|_\infty}{\|\hat{x}\|_\infty} = \|S_n\hat{x}\|_\infty = \|D_n\|_1.$$

Hence we conclude that $\|S_n\| = \|D_n\|_1$. ∎

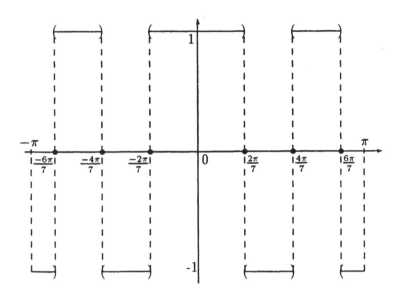

Figure 8.1: The function $\hat{x}$ for $n = 3$.

The next result says that if we consider $C_{2\pi}$ instead of $F_{2\pi}$ then the norm of the linear mapping S_n remains unchanged. The simple reason for this is that $C_{2\pi}$ sits inside $F_{2\pi}$ and somehow there are elements of $C_{2\pi}$ close to $\hat{x}$.

Lemma 8.20 *Let* $X = (C_{2\pi}, \|\cdot\|_\infty)$. *Then the operator* S_n *as an element of* $\mathcal{B}(X)$ *has norm* $\|D_n\|_1$.

Proof. As in **8.19** an application of **8.18** shows us that $\|S_n\| \leq \|D_n\|_1$. Take $\epsilon > 0$. We shall show that there is a function $\hat{y} \in C_{2\pi}$ such that $\|\hat{y}\|_\infty = 1$ and $\|S_n\hat{y}\|_\infty \geq \|D_n\|_1 - \epsilon$. This will then show that

$$\|S_n\| = \sup_{x \neq 0} \frac{\|S_n x\|_\infty}{\|x\|_\infty} \geq \frac{\|S_n\hat{y}\|_\infty}{\|\hat{y}\|_\infty} \geq \|D_n\|_1 - \epsilon,$$

and since this inequality holds for all $\epsilon > 0$, we must have $\|S_n\| \geq \|D_n\|_1$. All that remains is the construction of $\hat{y}$. It is clear that $\hat{y}$ should be close to the function $\hat{x}$, as constructed in **8.19**. More precisely, we ensure that

$$\|\hat{x} - \hat{y}\|_1 \leq \frac{\epsilon}{n + \frac{1}{2}}.$$

Once this is done, we will be able to reason as follows (making frequent use of **8.18**):

$$
\begin{aligned}
\|S_n \hat{y}\|_\infty &\geq |(S_n \hat{y})(0)| \\
&\geq |(S_n \hat{x})(0)| - |(S_n \hat{x})(0) - (S_n \hat{y})(0)| \\
&= \left| \|D_n\|_1 - \left| \frac{1}{2\pi} \int_{-\pi}^{\pi} [\hat{x}(s) - \hat{y}(s)] D_n(s) \, ds \right| \right. \\
&\geq \|D_n\|_1 - \frac{1}{2\pi} \int_{-\pi}^{\pi} |\hat{x}(s) - \hat{y}(s)||D_n(s)| \, ds \\
&\geq \|D_n\|_1 - \max_{s \in [-\pi, \pi]} |D_n(s)| \frac{1}{2\pi} \int_{-\pi}^{\pi} |\hat{x}(s) - \hat{y}(s)| \, ds \\
&= \|D_n\|_1 - \left(n + \frac{1}{2} \right) \|\hat{x} - \hat{y}\|_1 \\
&\geq \|D_n\|_1 - \epsilon.
\end{aligned}
$$

Now the construction of $\hat{y}$ is quite easy. It has to agree with $\hat{x}$ in value over 'most' of the interval $[-\pi, \pi]$, the quantification of 'most' being tied up with the norm $\| \cdot \|_1$ and the number $\epsilon/(n + 1/2)$. Denote the zeros of $\hat{x}$ by $t_{\pm k}$ so that

$$
t_{\pm k} = \pm \frac{k\pi}{n + 1/2}, \quad k = 1, 2, \ldots, n.
$$

For each point $t_{\pm k}$, define associated points $a_{\pm k}$ and $b_{\pm k}$ where $a_{\pm k} < t_{\pm k} < b_{\pm k}$ and

$$
|t_{\pm k} - a_{\pm k}| = |t_{\pm k} - b_{\pm k}| < \min \left(\frac{\pi}{2n + 1}, \frac{\pi\epsilon}{2n + 1} \right).
$$

By this device we ensure that the intervals $[a_{\pm k}, b_{\pm k}]$ do not overlap, and are small. Now set

$$
\hat{y}(t) = \frac{t - a_{\pm k}}{b_{\pm k} - a_{\pm k}} \hat{x}(b_{\pm k}) + \frac{b_{\pm k} - t}{b_{\pm k} - a_{\pm k}} \hat{x}(a_{\pm k}), \ t \in [a_{\pm k}, b_{\pm k}], \ k = 1, 2, \ldots
$$

and $\hat{y}(t) = \hat{x}(t)$ otherwise. This rather complex looking expression has the simple effect that $\hat{y}$ is made up of linear pieces as illustrated in 8.2 for $n = 3$. Notice that $\|\hat{y}\|_\infty = 1$ and so

$$
\begin{aligned}
\|\hat{x} - \hat{y}\|_1 &= \frac{1}{2\pi} \int_{-\pi}^{\pi} |\hat{x}(s) - \hat{y}(s)| \, ds \\
&= \frac{1}{2\pi} \sum_{k=1}^{n} \int_{a_{\pm k}}^{b_{\pm k}} |\hat{x}(s) - \hat{y}(s)| \, ds
\end{aligned}
$$

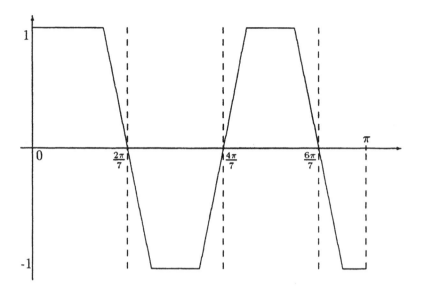

Figure 8.2: The function $\hat{y}$ for $n = 3$.

$$\leq \frac{1}{2\pi} \sum_{k=1}^{n} \int_{a_{\pm k}}^{b_{\pm k}} 2\,ds$$

$$= \frac{1}{2\pi} \sum_{k=1}^{n} 2|b_{\pm k} - a_{\pm k}|$$

$$\leq \frac{1}{2\pi} 4n \left(\frac{\pi \epsilon}{2n+1} \right)$$

$$< \frac{\epsilon}{n + 1/2}. \qquad \blacksquare$$

Lemma 8.21 *Let $X = (C_{2\pi}, \|\cdot\|_\infty)$. Then the linear mapping S_n in $\mathcal{B}(X)$ satisfies*

$$\|S_n\| \geq \frac{4}{\pi^2} \log n, \quad n \geq 1.$$

Proof. Once again the proof is highly manipulative. Since D_n is an even function on $[-\pi, \pi]$, we have

$$\|S_n\| = \|D_n\|_1 = \frac{1}{2\pi} \int_{-\pi}^{\pi} \left| \frac{\sin(n + \frac{1}{2})s}{\sin \frac{1}{2}s} \right| ds = \frac{1}{\pi} \int_0^{\pi} \left| \frac{\sin(n + \frac{1}{2})s}{\sin \frac{1}{2}s} \right| ds.$$

The change of variable $s = 2v$ and the inequality $\sin v \leq v$ for $v \geq 0$ gives

$$\|S_n\| = \frac{2}{\pi} \int_0^{\pi/2} \frac{|\sin(2n+1)v|}{\sin v} \, dv \geq \frac{2}{\pi} \int_0^{\pi/2} \frac{|\sin(2n+1)v|}{v} \, dv.$$

Another change of variable, $u = \pi v/(2n+1)$, gives

$$\|S_n\| \geq \frac{2}{\pi} \int_0^{n+1/2} \frac{|\sin \pi u|}{u} \, du > \frac{2}{\pi} \int_0^n \frac{|\sin \pi u|}{u} \, du.$$

Breaking the interval $[0, n]$ up into n subintervals, and using an obvious abuse of notation, we get

$$\begin{aligned}
\|S_n\| &> \frac{2}{\pi} \left(\int_0^1 + \int_1^2 + \ldots + \int_{n-1}^n \right) \frac{|\sin \pi u|}{u} \, du \\
&= \frac{2}{\pi} \int_0^1 \left(\frac{1}{u} + \frac{1}{u+1} + \ldots + \frac{1}{u+n-1} \right) \sin \pi u \, du \\
&\geq \frac{2}{\pi} \int_0^1 \left(1 + \frac{1}{2} + \ldots + \frac{1}{n} \right) \sin \pi u \, du.
\end{aligned}$$

Now it follows from exercise 4 on page 118 that

$$1 + \frac{1}{2} + \ldots + \frac{1}{n} > \log(n+1),$$

and so

$$\|S_n\| > \frac{2}{\pi} \log n \int_0^1 \sin \pi u \, du = \frac{4}{\pi^2} \log n. \quad \blacksquare$$

Now we are very close to our desired objective. The only point at issue before we can apply **8.16** is whether $(C_{2\pi}, \| \cdot \|_\infty)$ is complete. This cannot be a substantial difficulty, since this space resembles the space $(C[0, 1], \| \cdot \|_\infty)$ very closely.

Lemma 8.22 *The space $(C_{2\pi}, \| \cdot \|_\infty)$ is complete.*

Proof. Let $\{x_n\}$ be a Cauchy sequence in $C_{2\pi}$. Then the sequence $\{y_n\}$ in $C[0, 2\pi]$ obtained by restricting x_n to the interval $[0, 2\pi]$ (i.e., by taking $y_n(t) = x_n(t)$, $t \in [0, 2\pi]$) will be Cauchy in the space $(C[0, 2\pi], \| \cdot \|_\infty)$. Since this last space is complete (see **5.11**), $\{y_n\}$ converges to a function y in $C[0, 2\pi]$. Since $y_n(0) = y_n(2\pi)$ for $n \in \mathbb{N}$, we must have $y(0) = y(2\pi)$. Now extend y to a function $\tilde{y} \in C_{2\pi}$ by making $\tilde{y}$ 2π-periodic. Then $x_n \to \tilde{y}$ and so $(C_{2\pi}, \| \cdot \|_\infty)$ is complete. $\blacksquare$

Theorem 8.23 *There exists a function x in $C_{2\pi}$ such that the Fourier series of x does not converge uniformly to x.*

Proof. The statement of the theorem is couched in the language of classical analysis. In the jargon of abstract analysis, we would say that there is an x in $C_{2\pi}$ such that the partial sums of the Fourier series do not converge to x in the norm $\|\cdot\|_\infty$. This follows from **8.21** and the remarks following **8.16**. ■

Notice that **8.23** does not achieve the full strength of the Du Bois-Reymond result. It is possible to deduce this result from **8.23**, combined with some compactness arguments. The reader is refered to the exercises for an outline of this argument.

Exercises

1. Let $X = (C_{2\pi}, \|\cdot\|_\infty)$ and $Y = (C_{2\pi}, \|\cdot\|_1)$. Show that the Fourier partial sum operator S_0 has $\|S_0\| = 1$ when S_0 is regarded as a member of $\mathcal{B}(X)$ or $\mathcal{B}(Y)$.

2. Let $x \in C[0,1]$. Let $B_n x$ denote the nth Bernstein polynomial associated with x (see chapter 5). Show that the sequence $\{B_n\}$ is bounded, when $C[0,1]$ has the supremum norm.

3. Prove that there is a function $x \in C_{2\pi}$ and a point $t \in [0, 2\pi]$ such that the Fourier series for x diverges at the point t. This is the full strength of the Du Bois-Reymond result, and is established by making formal the following outline argument.
   (i) A look back over this chapter should reveal that there is an $x \in C_{2\pi}$ such that $\{\|S_n x\|_\infty\}$ is unbounded.
   (ii) Let $t_n \in [0, 2\pi]$ be a point such that $|(S_n x)(t_n)| = \|S_n x\|_\infty$ for $n = 0, 1, 2, \ldots$. A suitable compactness argument shows that some subsequence of $\{t_n\}$, say $\{u_n\}$ converges to $u \in [0, 2\pi]$.
   (iii) Show that $\{(S_n x)(u)\}$ cannot converge to $x(u)$.

4. Prove that, for any $n \in \mathbb{N}$,
$$1 + \frac{1}{2} + \frac{1}{3} + \ldots + \frac{1}{n} > \log(n + 1).$$

5. A linear combination of the functions
$$1, \ t \mapsto \cos t, \ \ldots t \mapsto \cos nt, \ t \mapsto \sin t, \ \ldots, t \mapsto \sin nt$$

is called a trigonometric polynomial of degree n. Show that $S_n m = m$ for all trigonometric polynomials m of degree n.

6. Let M be the set of all trigonometric polynomials, regarded as a subspace of $C_{2\pi}$. Show that M is dense in $C_{2\pi}$. [Hint: use the transformation $t = \cos\theta$ between $[0, \pi]$ and $[-1, 1]$ together with the Weierstrass result on the density of the algebraic polynomials in $C[-1, 1]$.]

9

Differentiation in $\mathbb{R}^2$

In chapter 5 we discovered that the polynomials were dense in the space $C[a, b]$, consisting of real-valued, continuous functions on the closed interval $[a, b]$. We also observed there that if knowledge of a particular function in $C[a, b]$ is required only to within a prescribed tolerance ϵ, then the function can be replaced by a polynomial without the user being able to discern that such a replacement has been carried out. Implicit in the discussion was the demand that the polynomial should 'model' the function closely *over the whole* of the interval $[a, b]$. This sort of approximation process is often called *global*. The contrasting situation is that of *local* approximation, where we seek knowledge of a particular function to a given tolerance over a small subset of the interval $[a, b]$. It is to be expected that using a global approximation for this purpose is going to be needlessly complicated and expensive.

The most common local approximation is via derivatives. If f in $C[a, b]$ is differentiable at a point $t_0 \in [a, b]$, then we can approximate f near t_0 by a straight line whose gradient is $f'(t_0)$ and which passes through the point $(t_0, f(t_0))$. The extent to which this approximation models the function depends on the way the derivative of the function f behaves in the neighbourhood of t_0. A slowly varying derivative will mean that the linear approximation is effective over quite a large region surrounding t_0, while a rapidly varying derivative will mean that high accuracy may only be achieved very close to t_0.

The purpose of this chapter is to discuss local approximation of real-valued functions by way of derivatives. However, we shall consider multivariate functions $f : \mathbb{R}^n \to \mathbb{R}$ where $n > 1$. At this point, the thrust of our discussion of normed spaces changes somewhat. Thus

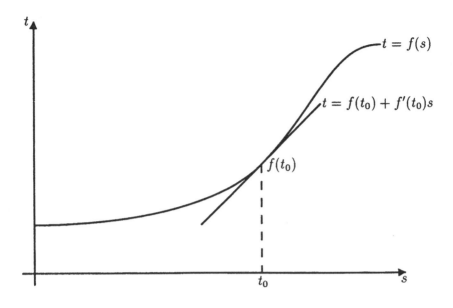

Figure 9.1: Approximation by the derivative.

far our attention has been focussed almost entirely on such spaces and their structure. Even when we considered linear mappings acting on such spaces, those linear mappings themselves formed a normed linear space, and so we were still studying an (albeit special) normed linear space. In this chapter, the importance of the linear space will diminish somewhat as we consider the action of mappings on the particular space $\mathbb{R}^n$.

Another decision must be made before we launch out into the general theory. We must decide whether to treat $\mathbb{R}^n$ for some general value of n or simply to discuss $\mathbb{R}^2$ and allow the reader to generalise the results to $\mathbb{R}^n$. We choose the latter course, since it avoids much of the notational complexity which would otherwise accompany our explanations. Every result in this chapter is stated only for functions from $\mathbb{R}^2$ to $\mathbb{R}$, but each has an obvious generalisation to functions from $\mathbb{R}^n$ to $\mathbb{R}$. The proofs also need only be altered in simple ways, and a good way for the reader to check her comprehension of a proof is to rewrite it in the more general case.

We begin with a short discussion of continuity. We will always assume, unless it is explicitly stated otherwise, that $\mathbb{R}^2$ is shorthand for the space $(\mathbb{R}^2, \| \cdot \|_\infty)$. Since the three norms $\| \cdot \|_\infty$, $\| \cdot \|_1$, $\| \cdot \|_2$ are equivalent (**2.14**), this choice of norm is simply a matter of convenience. Recall that in chapter 1 we investigated the continuity of functions from $\mathbb{R}^n$ to $\mathbb{R}$. Included amongst the results there were the following:

(i) the projection mappings defined by $(s, t) \mapsto s$ and $(s, t) \mapsto t$ are continuous

(ii) if f and g are continuous, real-valued functions on $\mathbb{R}^2$, then $f + g$ and $f.g$ are also continuous

(iii) if h is a continuous, real-valued function on $\mathbb{R}$ and f is a continuous, real-valued function on $\mathbb{R}^2$, then $h \circ f$ is a continuous, real-valued function on $\mathbb{R}^2$.

These facts can be used (as they were in chapter 2) to deduce the continuity of many functions on $\mathbb{R}^2$ which are defined by algebraic formulae. For example, they can be used to prove that the function defined by $f(s, t) = e^{st}$ is continuous on $\mathbb{R}^2$.

Now suppose f is a mapping from $\mathbb{R}^2$ to $\mathbb{R}$ (without necessarily assuming any continuity). It may happen that for each fixed s, $f(s, t)$ approaches some value dependent on s – call it $k(s)$ – as $t \to t_0$. It may also happen that $k(s)$ approaches some number l as $s \to s_0$. The number l is called the *repeated limit* and we would write

$$l = \lim_{s \to s_0} \lim_{t \to t_0} f(s, t).$$

As far as we are able to judge on the information available, there is every reason to suppose that the value of the repeated limit depends on the order of the individual limits. In general we expect

$$\lim_{s \to s_0} \lim_{t \to t_0} f(s, t) \neq \lim_{t \to t_0} \lim_{s \to s_0} f(s, t).$$

Example Take

$$f(s, t) = \frac{s^2 - t^2}{s^2 + t^2} \quad \text{for} \quad (s, t) \neq (0, 0).$$

For a fixed $s \neq 0$, $\lim_{t \to 0} f(s, t) = 1$ and so $\lim_{s \to 0} \lim_{t \to 0} f(s, t) = 1$. For a fixed $t \neq 0$, $\lim_{s \to 0} f(s, t) = -1$ and so $\lim_{t \to 0} \lim_{s \to 0} f(s, t)$ is -1. Thus it is easy to construct a function whose repeated limits are unequal.

Example Take

$$f(s,t) = \begin{cases} t\sin(1/s) & s \neq 0 \\ 0 & s = 0 \end{cases}.$$

For fixed $s \neq 0$, $\lim_{t\to 0} f(s,t) = 0$ and so $\lim_{s\to 0}\lim_{t\to 0} f(s,t) = 0$. For fixed $t \neq 0$, $\lim_{s\to 0} f(s,t)$ does not exist. Thus it is possible for one iterated limit to exist while the other does not. Notice also that the function f in this example is continuous at $(0,0)$ since

$$|f(s,t) - f(0,0)| = |f(s,t)| = |t\sin(1/s)| \leq |t|.$$

Thus if $\|(s,t)\|_\infty < \epsilon$, it follows that $|t| < \epsilon$, and so

$$|f(s,t) - f(0,0)| < \epsilon,$$

which demonstrates the continuity of f at $(0,0)$.

These simple examples are supposed to engender in us some caution about allowing processes to occur in individual coordinate directions. We can however state a positive result.

Lemma 9.1 *Suppose* $f : I\!R^2 \to I\!R$ *is continuous at* (s_0, t_0), *and* $f(s_0, t_0) = c$. *If for each* s *with* $0 < |s - s_0| < h$ *there is some number* $k(s)$ *such that* $\lim_{t\to t_0} f(s,t) = k(s)$, *then* $k(s) \to c$ *as* $s \to s_0$.

Proof. Take $\epsilon > 0$. Since f is continuous at (s_0, t_0) we can find $0 < \delta < h$ such that $|f(s,t) - f(s_0,t_0)| = |f(s,t) - c| < \epsilon/2$ whenever $|s - s_0| < \delta$ and $|t - t_0| < \delta$. Now for any s with $|s - s_0| < \delta$ we choose t such that $|t - t_0| < \delta$ and $|k(s) - f(s,t)| < \epsilon/2$. Then

$$\begin{aligned} |k(s) - c| &\leq |k(s) - f(s,t)| + |f(s,t) - c| \\ &< \frac{\epsilon}{2} + \frac{\epsilon}{2} \\ &= \epsilon. \end{aligned}$$

Hence $k(s) \to c$ as $s \to s_0$.   ∎

What **9.1** tells us is that it is impossible to find a repeated limit of a function f at a point of continuity of that function which differs from the value of the function at the point.

Having already remarked on the unreliability of information given in coordinate directions, it must be said that there is something very natural about such information. We now define the notion of a partial derivative, which is precisely another form of coordinate direction information. By now, our feeling should be that partial derivatives will not serve on their own as notions for a derivative in the sense of providing local approximation.

Definition 9.2 *The partial derivatives of* $f : \mathbb{R}^2 \to \mathbb{R}$ *at the point* (s_0, t_0) *are defined as*

$$(D_1 f)(s_0, t_0) = \lim_{h \to 0} \frac{1}{h} \{ f(s_0 + h, t_0) - f(s_0, t_0) \}$$

and

$$(D_2 f)(s_0, t_0) = \lim_{k \to 0} \frac{1}{k} \{ f(s_0, t_0 + k) - f(s_0, t_0) \}$$

whenever these limits exist.

Most readers will be aware of other notations for D_1 and D_2. For example $D_1 f$ is sometimes denoted by

$$D_s f, \quad f_s, \quad \frac{\partial f}{\partial s}.$$

These partial derivatives amount to 'fixing' one of the variables and differentiating the resulting univariate function in the other variable. Consequently, the theorems of univariate differential calculus apply to partial derivatives. For example, if $(D_1 f)(s, t_0)$ exists for all s in $[s_0, s_1]$, then there exists a point ξ in (s_0, s_1) such that

$$f(s_1, t_0) - f(s_0, t_0) = (D_1 f)(\xi, t_0)(s_1 - s_0).$$

This is the translation of the univariate mean value theorem to our present setting.

Example Take $f : \mathbb{R}^2 \to \mathbb{R}$ defined by

$$f(s, t) = \begin{cases} 0 & st = 0 \\ 1 & st \neq 0 \end{cases}.$$

It is easily seen that f is discontinuous at $(0, 0)$. However, $f(s, 0) = 0$ for all $s \in \mathbb{R}$ and $f(0, t) = 0$ for all $t \in \mathbb{R}$. Hence, $(D_1 f)(0, 0) = (D_2 f)(0, 0) = 0$. Thus although f is discontinuous at $(0, 0)$, the partial derivatives both exist at that point. This is in strong contrast to the univariate case, where the differentiability of a function at a point implied its continuity at that point.

We have been emphasizing so far that information about coordinate directions gives us little information about the local behaviour of a function. To get the appropriate generalisation of the univariate derivative, we must adopt a different approach. As is usual in mathematics, a deeper look at the univariate case will reveal structure that provides the appropriate basis on which to build our notion of a derivative for multivariate functions. Let f be a real-valued function on $\mathbb{R}$ which is differentiable at s with derivative A. Then

$$\lim_{h \to 0} \frac{f(s+h) - f(s)}{h} = A.$$

This means that given $\epsilon > 0$ there is an $\delta > 0$ such that

$$\left| \frac{f(s+h) - f(s)}{h} - A \right| < \epsilon \quad \text{whenever} \quad |h| < \delta.$$

This is equivalent to

$$|f(s+h) - f(s) - Ah| < \epsilon|h| \quad \text{whenever} \quad |h| < \delta.$$

If we think back to our earlier concept of the derivative being a good linear approximation to the function f near the point s, then we see that the last of the statements above describes just that situation. Remember that s is a fixed point and we are approximating $f(s+h)$ (the value of f near s) by the affine function g where $g(h) = f(s) + Ah$. The quantifiers ϵ and δ indicate that the quality of this approximation is good when $|h|$ is small. The word affine is used here to describe a function whose graph is a straight line but whose value at the origin is not necessarily zero, and we reserve the word linear for affine functions g such that $g(0) = 0$. The reason for this is that we want to keep in step with our use of the word linear in chapter 8. If f is a real-valued function on $\mathbb{R}^2$, then we want to construct our definition of the derivative in such a way that it gives an affine function of two variables (a plane geometrically), which approximates f in a neighbourhood of the specified point. Values of f in a neighbourhood of the point (s, t) can be expressed as $f(s+h, t+k)$ and so the affine function of h and k will be of the form $a + bh + ck$, where a, b and c are constants. When $h = k = 0$ we want the affine function to agree with f at (s, t), so this forces $a = f(s, t)$. The other constants are determined by demanding that the discrepancy between f and this affine function be small when we are near the point (s, t). Our natural measure of proximity is that $\|(h, k)\|_\infty$ should be small. Now a simple mimicking of the univariate case will suffice.

Definition 9.3 *Let f be a function from $\mathbb{R}^2$ to $\mathbb{R}$. Then f is differentiable at $(s, t) \in \mathbb{R}^2$ with gradient vector (A, B) if, given $\epsilon > 0$, there is a $\delta > 0$ such that*

$$|f(s + h, t + k) - f(s, t) - Ah - Bk| \le \epsilon \|(h, k)\|_\infty$$

whenever $\|(h, k)\|_\infty \le \delta$.

The important question which confronts us now is whether the numbers A and B can be easily calculated in a given situation. The next result shows that they can.

Lemma 9.4 *If $f : \mathbb{R}^2 \to \mathbb{R}$ is differentiable at (s, t) with gradient vector (A, B) then $A = (D_1 f)(s, t)$ and $B = (D_2 f)(s, t)$.*

Proof. Suppose f is differentiable at (s, t) with gradient vector (A, B). This means that, in particular, given $\epsilon > 0$ there is a $\delta > 0$ such that

$$|f(s + h, t + 0) - f(s, t) - Ah - B.0| \le \epsilon \|(h, 0)\|_\infty = \epsilon |h|,$$

whenever $\|(h, 0)\|_\infty = |h| \le \delta$. Here we have applied the definition using the 'incremental' vector $(h, 0)$. Dividing by $|h|$, we obtain

$$\left| \frac{f(s + h, t) - f(s, t)}{h} - A \right| \le \epsilon \quad \text{whenever} \quad 0 < |h| < \delta.$$

Thus

$$\lim_{h \to 0} \frac{f(s + h, t) - f(s, t)}{h} = A,$$

and so $(D_1 f)(s, t)$ exists and has value A. The result $B = (D_2 f)(s, t)$ follows similarly. ∎

Care must be taken not to read **9.4** incorrectly. This result does *not* say that if $D_1 f$ and $D_2 f$ exist at some point, then these are the components of the gradient vector. We have to know that the gradient vector exists *before* we can conclude that its components are the partial derivatives of f. The following example illustrates the sort of use to which **9.4** is put.

Example The function $f : \mathbb{R}^2 \to \mathbb{R}$ defined by $f(s,t) = st$ is differentiable at each point $(s,t) \in \mathbb{R}^2$ with gradient vector (t,s). This can be established as follows. Suppose we knew in advance that f was differentiable at (s,t). Then by **9.4** its gradient vector would have to be (t,s). Consequently, to check whether f *is* differentiable at (s,t) we have to check whether, given $\epsilon > 0$, there is a $\delta > 0$ such that

$$|f(s+h, t+k) - f(s,t) - th - sk| < \epsilon \|(h,k)\|_\infty$$

whenever $\|(h,k)\|_\infty < \delta$. This follows easily from the following calculation:

$$|f(s+h, t+k) - f(s,t) - th - sk| = |(s+h)(t+k) - st - th - sk| = |hk|.$$

Now taking $|h|, |k| \leq \min\{1, \epsilon\}$ gives $|hk| \leq \epsilon$, and so for such values of (h,k),

$$|f(s+h, t+k) - f(s,t) - th - sk| \leq \epsilon \|(h,k)\|_\infty.$$

The following set of exercises explore some of the elementary properties of the derivative.

Exercises

1. What are the iterated limits of

$$\frac{\sin^2 s}{s^2 + 2t^2} \quad \text{and} \quad \frac{s^2}{s^2 + t^2 + 1}$$

   at $(0,0)$?

2. The following is an exercise on iterated limits for series. Suppose $a_{mn} \geq 0$ for all $m, n \in \mathbb{N}$. Suppose that $\sum_{n=1}^\infty a_{mn} = S_m$, $m \in \mathbb{N}$. Prove that $\sum_{m=1}^\infty S_m$ converges. Let $\sum_{m=1}^\infty S_m = S$, and $\sum_{m=1}^\infty a_{mn} = T_n$. Show that $\sum_{n=1}^\infty T_n \leq S$. Deduce that $\sum_{n=1}^\infty T_n = \sum_{m=1}^\infty S_m$.

3. If $f_1, f_2 : \mathbb{R}^2 \to \mathbb{R}$ are differentiable at (s,t) with gradient vectors (A_1, B_1) and (A_2, B_2) respectively, then $f_1 + f_2$ is differentiable at (s,t) with gradient vector $(A_1 + A_2, B_1 + B_2)$. Also αf_1, $\alpha \in \mathbb{R}$, is differentiable with gradient vector $\alpha(A_1, B_1)$. (That is, differentiation is a linear process.)

4. If f and g are real-valued functions on $\mathbb{R}$ which are differentiable at s_0 and t_0 respectively, then the function $F : \mathbb{R}^2 \to \mathbb{R}$ defined by $F(s,t) = f(s)g(t)$ is differentiable at (s_0, t_0), with gradient

$$(f'(s_0)g(t_0), f(s_0)g'(t_0)).$$

5. If $f : \mathbb{R}^2 \to \mathbb{R}$ is defined by $f(s,t) = \alpha s + \beta t + \gamma$, $\alpha, \beta, \gamma \in \mathbb{R}$, then f is differentiable for all $(s,t) \in \mathbb{R}^2$ and has gradient vector (α, β). Thus a plane is its own derivative function.

6. Let $f : \mathbb{R}^2 \to \mathbb{R}$ be defined by $f(s,t) = (st)^{\frac{1}{3}}$. Prove that
   (i) f is continuous at $(0,0)$
   (ii) $(D_1 f)(0,0) = (D_2 f)(0,0) = 0$
   (iii) f is *not* differentiable at $(0,0)$.
   Thus continuity at a point does not guarantee differentiablity there.

Lemma 9.5 *If $f : \mathbb{R}^2 \to \mathbb{R}$ is differentiable at (s,t), then f is continuous at that point.*

Proof. Suppose f has gradient vector (A, B) at (s,t). Take $\epsilon > 0$ and choose $\delta > 0$ so that

$$|f(s+h, t+k) - f(s,t) - Ah - Bk| < \|(h,k)\|_\infty$$

whenever $\|(h,k)\|_\infty < \delta$. (Thus we have taken the value of ϵ in **9.3** as 1.) Now take $\delta' > 0$ so that

$$\delta' = \min\left(\delta, \frac{\epsilon}{|A| + |B| + 1}\right).$$

Whenever $\|(h,k)\|_\infty < \delta'$ we have

$$
\begin{aligned}
|f(s+h, t+k) - f(s,t)| \;&\leq\; |f(s+h, t+k) - f(s,t) \\
&\qquad - Ah - Bk| + |Ah + Bk| \\
&\leq\; \|(h,k)\|_\infty + |Ah| + |Bk| \\
&\leq\; \delta' + |A|\|(h,k)\|_\infty + |B|\|(h,k)\|_\infty \\
&\leq\; \delta'(1 + |A| + |B|) \\
&\leq\; \epsilon.
\end{aligned}
$$

Hence f is continuous at (s,t). ∎

We have already seen that the existence of the partial derivatives of a function $f : \mathbb{R}^2 \to \mathbb{R}$ at a given point is not sufficient to guarantee the existence of the derivative of f at that point. However, it is possible to detect whether f is differentiable if slightly more information about the partial derivatives is available.

Lemma 9.6 *Let f be a real-valued function on $\mathbb{R}^2$ and let (s,t) be a point in $\mathbb{R}^2$. Suppose there exists a $\delta > 0$ such that $D_1 f$ exists in $B_\delta((s,t))$ and is continuous at (s,t). Suppose also $D_2 f$ exists at (s,t). Then f is differentiable at (s,t).*

Proof. Let $(D_1 f)(s,t) = A$ and $(D_2 f)(s,t) = B$. Take $\epsilon > 0$. We have to show that there is a $\delta > 0$ such that

$$|f(s+h,t+k) - f(s,t) - Ah - Bk| < \epsilon \|(h,k)\|_\infty$$

whenever $\|(h,k)\|_\infty < \delta$. We first use the continuity of $D_1 f$ at (s,t) to assert that there is a $\delta_1 > 0$ such that

$$|(D_1 f)(s',t') - A| \leq \epsilon/2 \quad \text{whenever} \quad \|(s-s',t-t')\|_\infty < \delta_1.$$

Then, since $(D_2 f)(s,t) = B$, there is a $\delta_2 > 0$ such that

$$|f(s,t+k) - f(s,t) - Bk| < \epsilon|k|/2$$

whenever $|k| < \delta_2$. Set $\delta_0 = \min(\delta_1, \delta_2)$ and take (h,k) such that $\|(h,k)\|_\infty < \delta$, that is $|h| < \delta$ and $|k| < \delta$. Then, since $D_1 f$ exists in $B_\delta((s,t))$, we may apply the one-dimensional mean-value theorem to obtain

$$f(s+h,t+k) - f(s,t+k) = h(D_1 f)(s',t+k)$$

for some s' in $(s, s+h)$. Finally, with the same restrictions on h and k,

$$
\begin{aligned}
|f(s+h,&t+k) - f(s,t) - Ah - Bk| \\
&\leq |f(s+h,t+k) - f(s,t+k) - Ah| \\
&\quad + |f(s,t+k) - f(s,t) - Bk| \\
&< |f(s+h,t+k) - f(s,t+k) - h(D_1 f)(s',t+k)| \\
&\quad + |h(D_1 f)(s',t+k) - Ah| + \epsilon|k|/2 \\
&= |h(D_1 f)(s',t+k) - Ah| + \epsilon|k|/2 \\
&< \epsilon|h|/2 + \epsilon|k|/2 \\
&\leq \epsilon\|(h,k)\|_\infty. \qquad \blacksquare
\end{aligned}
$$

In univariate calculus, one of the great benefits of the theory is the ease with which maxima and minima of functions can be detected. We start now on a series of results which will lead us to a criterion for the detection of the maxima and minima of a function $f : \mathbb{R}^2 \to \mathbb{R}$. For such a function, a point (s_0, t_0) will be called a *relative minimum* of f if there exists $\delta > 0$ such that $f(s_0, t_0) \le f(s, t)$ for all (s, t) in $B_\delta((s_0, t_0))$. A *relative maximum* is defined similarly.

Lemma 9.7 *Suppose* $f : \mathbb{R}^2 \to \mathbb{R}$ *has a relative maximum or minimum at* (s, t) *in* $\mathbb{R}^2$. *Suppose also that* $D_1 f$ *and* $D_2 f$ *exist at* (s, t). *Then* $(D_1 f)(s, t) = (D_2 f)(s, t) = 0$.

Proof. Suppose f has a relative minimum at (s, t). Recall that

$$(D_1 f)(s, t) = \lim_{h \to 0} \frac{1}{h} \{ f(s + h, t) - f(s, t) \}.$$

Take $\delta > 0$ such that $f(s + h, t + k) > f(s, t)$ for all $\|(h, k)\|_\infty < \delta$. For any $\delta > h > 0$ we have

$$\frac{1}{h} \{ f(s + h, t) - f(s, t) \} \ge 0,$$

while for any $-\delta < h < 0$ we have

$$\frac{1}{h} \{ f(s + h, t) - f(s, t) \} \le 0.$$

This forces $(D_1 f)(s, t) = 0$. The other case follows similarly. ∎

Corollary 9.8 *Suppose* $f : \mathbb{R}^2 \to \mathbb{R}$ *has a relative minimum or maximum at* $(s, t) \in \mathbb{R}^2$. *If* f *is differentiable at* (s, t), *then its gradient vector is* $(0, 0)$.

Proof. From **9.4** the gradient vector is

$$((D_1 f)(s, t), (D_2 f)(s, t))$$

and the result follows from an application of **9.7**. ∎

If we think of the analogous situation in univariate calculus, then we do not expect **9.7** to be sufficient to guarantee the existence of a relative maximum or minimum. The following set of exercises show just how wide the variety of behaviour can be even though the conditions of **9.7** are met.

Exercises

1. Let $f : \mathbb{R}^2 \to \mathbb{R}$ be defined by $f(s,t) = s^2t^2$. Show that $(D_1f)(0,0) = (D_2f)(0,0) = 0$ and that f has a relative minimum at $(0,0)$. Sketch the surface in $\mathbb{R}^2$ which represents f.

2. Let $f : \mathbb{R}^2 \to \mathbb{R}$ be defined by $f(s,t) = st$. Show that $(D_1f)(0,0)$ and $(D_2f)(0,0)$ are both 0. Show also that

$$f(0,0) \; < \; f(s,t) \quad \text{for} \quad st > 0$$
$$f(0,0) \; > \; f(s,t) \quad \text{for} \quad st < 0.$$

Thus f does not have a local maximum or minimum at $(0,0)$. Sketch the surface in $\mathbb{R}^3$ which represents f. The point $(0,0)$ in this example is sometimes called a *saddle point*.

3. Let $f : \mathbb{R}^2 \to \mathbb{R}$ be defined by $f(s,t) = (t - s^2)(t - 2s^2)$. Show that $(D_1f)(0,0) = (D_2f)(0,0) = 0$. Define $g_\lambda : \mathbb{R} \to \mathbb{R}$ by

$$g_\lambda(s) = (\lambda s - s^2)(\lambda s - 2s^2).$$

Then g_λ is the restriction of the function f to the line $t = \lambda s$ in $\mathbb{R}^2$. Show that for all $\lambda \in \mathbb{R}$, g_λ has a local minimum at the origin. Show also that for any $\delta > 0$ the ball $B_\delta(\theta)$ contains some points where f is strictly positive and others where f is strictly negative, so that f cannot have a relative minimum at $\theta = (0,0)$.

4. Define $f : \mathbb{R}^2 \to \mathbb{R}$ by $f(s,t) = s^2|t|$. Show that f has a relative minimum at $(0,0)$, but that f is not differentiable at $(0,0)$.

If we visualise $f : \mathbb{R}^2 \to \mathbb{R}$ as defining a surface over $\mathbb{R}^2$ then f can have a line of relative maxima, a so-called 'saddle point', a line of 'points of inflection', and so on. Figure 9.2 illustrates a function which has a local maximum, while Figure 9.3 illustrates one which has a saddle point. In each of these illustrations the first partial derivatives at the indicated points are zero, and our aim is to distinguish between maxima/minima and other forms of behaviour. The first result we need is called the *chain rule*. In the next chapter a much more general version of this result will be given.

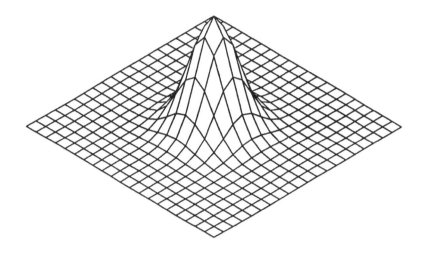

Figure 9.2: A function with a local maximum.

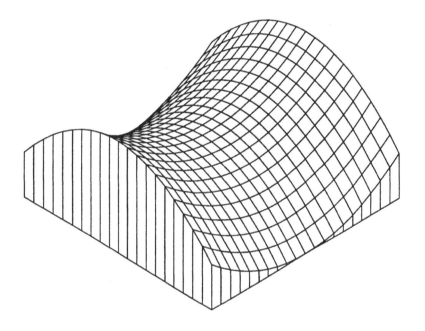

Figure 9.3: A function with a saddle point

Lemma 9.9 *Suppose* $f : \mathbb{R}^2 \rightarrow \mathbb{R}$ *is differentiable at all the points* $z_r = (s + rh, t + rk)$ *where* (s, t) *and* (h, k) *are fixed points in* $\mathbb{R}^2$ *and* $0 \leq r \leq 1$. *Define* $F : [0, 1] \rightarrow \mathbb{R}$ *by*

$$F(r) = f(z_r) = f(s + rh, y + rk).$$

Then

$$F'(r) = h(D_1 f)(z_r) + k(D_2 f)(z_r), \quad 0 < r < 1.$$

Proof. Fix r in $(0, 1)$ and take $\epsilon > 0$. Since f is differentiable at z_r there is a $\delta > 0$ such that

$$|f(s + r'h, t + r'k) - f(s + rh, t + rk) - (r' - r)h(D_1 f)(z_r)$$
$$-(r' - r)k(D_2 f)(z_r)| \leq \epsilon\|(\{r' - r\}h, \{r' - r\}k)\|_\infty$$

whenever $\|(\{r' - r\}h, \{r' - r\}k)\|_\infty < \delta$. Now take $\delta_1 > 0$ such that $\delta_1|h| < \delta$ and $\delta_1|k| < \delta$. Then whenever $|r' - r| < \delta_1$, we will have

$$\|(\{r' - r\}h, \{r' - r\}k)\|_\infty = \max\{|r' - r|h, |r' - r|k\}$$
$$\leq \max(\delta_1|h|, \delta_1|k|)$$
$$< \delta.$$

Suppose $|r' - r| < \delta_1$. Then

$$\left| \frac{F(r') - F(r)}{r' - r} - h(D_1 f)(z_r) - k(D_2 f)(z_r) \right| =$$

$$= \frac{1}{|r' - r|}|f(s + r'h, t + r'k) - f(s + rh, t + rk)$$
$$-(r' - r)h(D_1 f)(z_r) - (r' - r)k(D_2 f)(z_r)|$$
$$\leq \frac{\epsilon}{|r' - r|}\|(\{r' - r\}h, \{r' - r\}k)\|_\infty$$
$$\leq \epsilon\|(h, k)\|_\infty.$$

This shows that

$$\lim_{r' \rightarrow r} \frac{F(r') - F(r)}{r' - r} = h(D_1 f)(z_r) + k(D_2 f)(z_r),$$

which is what is required. ■

The characterisation of maxima and minima in the univariate case involved the first and second derivatives of the function. Briefly, that characterisation could be deduced as follows. Let $g : \mathbb{R} \to \mathbb{R}$ be a function which is twice differentiable in some interval $(a - \delta, a + \delta)$ where $\delta > 0$. Then Taylor's theorem says that for any h with $|h| < \delta$

$$g(a + h) = g(a) + hg'(a) + \frac{h^2}{2}g''(\xi_h),$$

where ξ_h is some point lying between $s = a$ and $s = a + h$. Now suppose $g'(a) = 0$ and $g''(s) > 0$ for $s \in (a - \delta, a + \delta)$. Then for $|h| < \delta$,

$$g(a + h) - g(a) = 0 + \frac{h^2}{2}g''(\xi_h) > 0.$$

This means that g has a relative minimum at $s = a$. In order to generalise this result, we have to seek the analogue of Taylor's theorem for functions $f : \mathbb{R}^2 \to \mathbb{R}$. This involves generalising the notion of higher order derivatives, which turns out to be a simple process.

If f is a mapping from $\mathbb{R}^2$ into $\mathbb{R}$, then it may have two partial derivatives $D_1 f$ and $D_2 f$. Each of these is a mapping with domain in $\mathbb{R}^2$ and range in $\mathbb{R}$, and so each in turn may also have two partial derivatives. We denote the two partial derivatives of $D_1 f$ by $D_1(D_1 f)$ and $D_2(D_1 f)$. The two partial derivatives of $D_2 f$ would be written $D_1(D_2 f)$ and $D_2(D_2 f)$. The notation is invariably streamlined so that $D_{ij} f$ represents $D_i(D_j f)$, $1 \leq i, j \leq 2$. There are again alternative ways of writing $D_{ij} f$. For example,

$$D_{12} f = \frac{\partial^2 f}{\partial s \partial t} = f_{st}.$$

What is more important is that these three sets of symbols all represent the same process, which is that of first taking the partial derivative of f in the t-variable, and then taking the partial derivative of this new function with respect to the s-variable. There is no reason to suppose, for a given f, that $D_{12} f = D_{21} f$, and the exercises contain examples of functions for which this equality is not true. We could establish relatively simple conditions for this statement to be true, but we defer this until the chapter on Riemann integration – when the desired result will be a particularly nice consequence of that theory. Instead we continue our attempt to describe relative maxima or minima. The following theorem contains the analogue of the mean-value theorem for univariate calculus.

Theorem 9.10 *Let* $z_r = (s+rh, t+rk)$, $0 \le r \le 1$, *be a line segment in* $\mathbb{R}^2$.

(i) Suppose $f : \mathbb{R}^2 \to \mathbb{R}$ *is differentiable at each point* z_r, $0 \le r \le 1$. *Then there is an* r', $0 < r' < 1$, *such that*

$$f(z_1) - f(z_0) = h(D_1 f)(z_{r'}) + k(D_2 f)(z_{r'}).$$

(ii) Suppose $f : \mathbb{R}^2 \to \mathbb{R}$ *is such that* $D_1 f$ *and* $D_2 f$ *are both differentiable at each point* z_r, $0 \le r \le 1$. *Then there is an* r', $0 < r' < 1$ *such that*

$$
\begin{aligned}
f(z_1) - f(z_0) &= h(D_1 f)(z_0) + k(D_2 f)(z_0) + \frac{1}{2} h^2 (D_{11} f)(z_{r'}) \\
&+ \frac{1}{2} hk (D_{12} f)(z_{r'}) + \frac{1}{2} hk (D_{21} f)(z_{r'}) + \frac{1}{2} k^2 (D_{22} f)(z_{r'}).
\end{aligned}
$$

Proof. (i) Define the function $F : [0, 1] \to \mathbb{R}$ by

$$F(r) = f(z_r) = f(s + rh, t + rk), \quad 0 < r < 1.$$

By **9.9**,

$$F'(r) = h(D_1 f)(z_r) + k(D_2 f)(z_r), \quad 0 < r < 1.$$

Now by the usual univariate mean value theorem, there is an $r' \in (0, 1)$ such that

$$F(1) - F(0) = F'(r')(1 - 0) = F'(r'),$$

that is,

$$f(z_1) - f(z_0) = h(D_1 f)(z_{r'}) + k(D_2 f)(z_{r'}).$$

(ii) Applying Taylor's theorem to the univariate function F defined above, we have that

$$F(1) = F(0) + F'(0) + \frac{1}{2} F''(r') \quad \text{for some } 0 < r' < 1.$$

From (i)

$$
\begin{aligned}
F'(r) &= h(D_1 f)(z_r) + k(D_2 f)(z_r), \quad 0 < r < 1 \\
&= h(D_1 f)(s + rh, t + rk) + k(D_2 f)(s + rh, t + rk).
\end{aligned}
$$

Applying **9.9** again, we obtain

$$F''(r) = h^2 (D_{11} f)(z_r) + hk (D_{12} f)(z_r) + hk (D_{21} f)(z_r) + k^2 (D_{22} f)(z_r),$$

for $0 < r < 1$. Substituting these two results in the formula for $F(1)$ obtained from Taylor's theorem gives the required result. ∎

This result can be made to look like the traditional univariate theorem if we adopt the correct notation. Firstly, we will use Df(a) to denote the gradient vector of the function f at the point $a \in \mathbb{R}^2$. Secondly, if $a = (s,t)$ and $b = (s',t')$ we will use ab to denote the usual scalar product $ss' + tt'$. The following is now a restatement of **9.10(i)**.

Theorem 9.11 *Let* $a,b \in \mathbb{R}^2$ *and suppose* $f : \mathbb{R}^2 \to \mathbb{R}$ *is differentiable at every point on the line segment joining* a *and* b. *Then there is a point* c *on that same line segment, such that*

$$f(b) - f(a) = Df(c)(b - a).$$

Proof. To see how the restatement takes place, we set $a = (s,t)$ and $b = (s+h, t+k)$. Thus in the notation of **9.10**, $a = z_0$ and $b = z_1$, so that $f(b) - f(a) = f(z_1) - f(z_0)$. Also

$$
\begin{aligned}
h(D_1f)(z_{r'}) + k(D_2f)(z_{r'}) &= (h,k)((D_1f)(z_{r'}), (D_2f)(z_{r'})) \\
&= (h,k)Df(z_{r'}) \\
&= (z_1 - z_0)Df(z_{r'}).
\end{aligned}
$$

This shows that, in the notation of **9.10**,

$$f(z_1) - f(z_0) = Df(z_{r'})(z_1 - z_0),$$

or, in the present notation,

$$f(b) - f(a) = Df(c)(b - a). \qquad \blacksquare$$

The reader is guided in the exercises through the generalisation of **9.10(ii)**. This will serve to demonstrate the beauty and power of the edifice we are constructing, since Taylor's theorem with two terms (**9.10(ii)**) will resemble the usual univariate result very closely.

In the following discussion, leading up to sufficient conditions for relative maxima and minima, we shall suppose that $f : \mathbb{R}^2 \to \mathbb{R}$ is such that D_1f and D_2f are differentiable at every point on the line segment joining two given points z_1 and z_0. This allows us to infer from **9.10(ii)** that there is a point on this line segment such that, if $z_1 - z_0 = (h, k)$, then

$$
\begin{aligned}
f(z_1) - f(z_0) &= h(D_1f)(z_0) + k(D_2f)(z_0) + \frac{1}{2}h^2(D_{11}f)(z_{r'}) \\
&+ \frac{1}{2}hk(D_{12}f)(z_{r'}) + \frac{1}{2}hk(D_{21}f)(z_{r'}) + \frac{1}{2}k^2(D_{22}f)(z_{r'}).
\end{aligned}
$$

Now suppose we are looking at a point z_0 which we suspect is a relative maximum or minimum. It must be that $(D_1 f)(z_0) = (D_2 f)(z_0) = 0$. Suppose also that for the function f, $D_{12} f = D_{21} f$. Then we will obtain

$$f(z_1) - f(z_0) = \frac{1}{2} h^2 (D_{11} f)(z_{r'}) + hk(D_{12} f)(z_{r'}) + \frac{1}{2} k^2 (D_{22} f)(z_{r'}). \quad (\dagger)$$

A relative maximum will have $f(z_1) - f(z_0) < 0$ for all z_1 sufficiently close to z_0, and so we need to examine the sign of the expression on the right-hand side of the above inequality. It is a quadratic expression in h and k of the form

$$Ah^2 + 2Bhk + Ck^2.$$

If $A \neq 0$, then we can complete the square to get

$$Ah^2 + 2Bhk + Ck^2 = \frac{1}{A} \left[(Ah + Bk)^2 + (AC - B^2)k^2 \right].$$

If $AC - B^2 > 0$ then certainly $A \neq 0$ and the sign of our quadratic expression is controlled completely by the sign of A. If $A < 0$ then the quadratic expression will be negative, and we will have

$$f(z_1) - f(z_0) < 0,$$

while if $A > 0$ we will obtain

$$f(z_1) - f(z_0) > 0.$$

We now formalise this discussion. It will be helpful to denote the line segment joining z_0 and z_1 by (z_0, z_1) so that

$$(z_0, z_1) = \{ \lambda z_0 + (1 - \lambda) z_1 : \lambda \in [0, 1] \}.$$

Theorem 9.12 *Let* $f : \mathbb{R}^2 \to \mathbb{R}$ *have continuous second partial derivatives in* $B_\delta(z_0)$, *where* $z_0 \in \mathbb{R}^2$ *and* $\delta > 0$. *Suppose also that* $(D_1 f)(z_0) = (D_2 f)(z_0) = 0$. *Set*

$$\Delta = (D_{11} f)(z_0)(D_{22} f)(z_0) - [(D_{12} f)(z_0)]^2.$$

(i) If $\Delta > 0$ *and if* $(D_{11} f)(z_0) > 0$, *then* f *has a relative minimum at* z_0.

(ii) If $\Delta > 0$ *and if* $(D_{11} f)(z_0) < 0$, *then* f *has a relative maximum at* z_0.

Proof. A result (11.8) from chapter **11** shows that $(D_{12}f)(z) = (D_{21}f)(z)$ for all $z \in B_\delta(z_0)$. Now suppose that $\Delta > 0$. Since the second partial derivatives are continuous by hypothesis, there is an ϵ, $0 < \epsilon < \delta$, such that

$$\Gamma(z) = (D_{11}f)(z)(D_{22}f)(z) - [(D_{12}f)(z)]^2 > 0$$

and $(D_{11}f)(z) \neq 0$ for all $z \in B_\epsilon(z_0)$. From the argument preceeding the theorem we obtain, for all $z_1 \in B_\epsilon(z_0)$,

$$f(z_1) - f(z_0) = \frac{1}{(D_{11}f)(z)}[\{h(D_{11}f)(z) + k(D_{12}f)(z)\}^2 + \Gamma(z)k^2],$$

for $z \in (z_1, z_0)$. It now follows immediately that if $(D_{11}f)(z) > 0$ for $z \in B_\epsilon(z_0)$ then

$$f(z_1) - f(z_0) > 0 \quad \text{for all } z_1 \in B_\epsilon(z_0),$$

while if $(D_{11}f)(z) < 0$ for $z \in B_\epsilon(z_0)$ then

$$f(z_1) - f(z_0) < 0 \quad \text{for all } z_1 \in B_\epsilon(z_0).$$

These are precisely the conditions for a relative minimum and maximum respectively. ∎

The idea of a 'saddle point' was mentioned in the last set of exercises. The precise meaning of this term is that $f : \mathbb{R}^2 \to \mathbb{R}$ has a *saddle* point at z_0, if $(D_1f)(z_0)$ and $(D_2f)(z_0)$ are both zero and if in every ball $B_\delta(z_0)$, $\delta > 0$, there exists a point z at which $f(z)$ is strictly greater than $f(z_0)$ and a point y such that $f(y)$ is strictly less than $f(z_0)$. This definition enables us to obtain a clearer picture of what happens in the case $\Delta < 0$.

Theorem 9.13 *Let* $f : \mathbb{R}^2 \to \mathbb{R}$ *have continuous second partial derivatives in* $B_\delta(z_0)$, *where* $z_0 \in \mathbb{R}^2$ *and* $\delta > 0$. *Suppose that both* $(D_1f)(z_0)$ *and* $(D_2f)(z_0)$ *are zero. If*

$$\Delta = (D_{11}f)(z_0)(D_{22}f)(z_0) - [(D_{12}f)(z_0)]^2 < 0,$$

then f *has a saddle point at* z_0.

Proof. As in **9.12** we may assume that $(D_{12}f)(z) = (D_{21}f)(z)$ for all $z \in B_\delta(z_0)$. There are now 3 cases.

Case (i). $(D_{11}f)(z_0) \neq 0$. In this case choose $0 < \epsilon < \delta$ so that

$$\Gamma(z) = (D_{11}f)(z)(D_{22}f)(z) - [(D_{12}f)(z)]^2 < 0$$

and $(D_{11}f)(z) \neq 0$ for all $z \in B_\epsilon(z_0)$. As before, for all $z_1 \in B_\epsilon(z_0)$,

$$f(z_1) - f(z_0) = \frac{1}{(D_{11}f)(z)}[\{h(D_{11}f)(z) + k(D_{12}f)(z)\}^2 + \Gamma(z)k^2].$$

Now let $B_r(z_0)$, $r > 0$, be any ball centred on z_0. Take $(h, 0)$ in $B_r(z_0) \cap B_\epsilon(z_0)$. Then the sign of $f(z_1) - f(z_0)$ is the same as the sign of $(D_{11}f)(z_0)$. On the other hand, take $k \neq 0$ such that

$$h = -\frac{(D_{12}f)(z_0)}{(D_{11}f)(z_0)}k.$$

Restrict h further if necessary, so that $(h, k) \in B_r(z_0) \cap B_\epsilon(z_0)$. Then

$$f(z_1) - f(z_0) = \frac{k^2}{(D_{11}f)(z)}\left[\left\{-\frac{(D_{11}f)(z)(D_{12}f)(z_0)}{(D_{11}f)(z_0)} + (D_{12}f)(z)\right\}^2 + \Gamma(z)\right].$$

Now as $k \to 0$, $z \to z_0$, and so, by the continuity of the second partial derivatives,

$$-\frac{(D_{11}f)(z)(D_{12}f)(z_0)}{(D_{11}f)(z_0)} + (D_{12}f)(z) \to 0 \quad \text{as} \quad k \to 0.$$

Since $\Gamma(z_0) < 0$, the expression for $f(z_1) - f(z_0)$ will take on the opposite sign to $(D_{11}f)(z_0)$ for sufficiently small values of k (i.e., for z_1 sufficiently close to z_0).

Case (ii). $(D_{22}f)(z_0) \neq 0$. We write

$$f(z_1) - f(z_0) = \frac{1}{(D_{22}f)(z)}[\{k(D_{22}f)(z) + h(D_{12}f)(z)\}^2 + \Gamma(z)h^2].$$

Now the argument is similar to case (i), with the roles of h and k and those of $D_{11}f$ and $D_{22}f$ reversed.

Case (iii) $(D_{11}f)(z_0) = (D_{22}f)(z_0) = 0$. We deduce from this assumption, and the hypothesis of our theorem, that $(D_{12}f)(z_0) \neq 0$. From the equation (†) on page 137 with $h = k$,

$$f(z_1) - f(z_0) = \frac{k^2}{2}[(D_{11}f)(z) + 2(D_{12}f)(z) + (D_{22}f)(z)].$$

Again by the continuity of the second partial derivatives, we have $(D_{11}f)(z) \rightarrow 0$ and $(D_{22}f)(z) \rightarrow 0$ as $z \rightarrow z_0$. Thus for z_1 sufficiently close to z_0, the sign of $f(z_1) - f(z_0)$ will be the same as that of $(D_{12}f)(z_0)$. The same argument as (i) with $h = -k$ produces points z_1 close to z_0 such that the sign of $f(z_1) - f(z_0)$ is the same as that of $-(D_{12}f)(z_0)$. ■

Exercises

1. (i) If $f : \mathbb{R}^2 \rightarrow \mathbb{R}$ is such that $|f(s,t)| \leq |st|$ for all $(s,t) \in \mathbb{R}^2$, prove that f is differentiable at $(0,0)$. Was it necessary to have the condition on f holding for all $(s,t) \in \mathbb{R}^2$?
   (ii) Let $f : \mathbb{R}^2 \rightarrow \mathbb{R}$ be defined by $f(s,t) = s|t|$. At which points in $\mathbb{R}^2$ is f differentiable?

2. Let $A = [a,b] \times [c,d]$, where $a, b, c, d \in \mathbb{R}$.
   (i) If $(D_2f)(z) = 0$ for all $z \in A$, show that there is a function $\phi : \mathbb{R} \rightarrow \mathbb{R}$ such that $f(s,t) = \phi(s)$, $(s,t) \in A$.
   (ii) If $(D_{12}f)(z) = 0$ for all $z \in A$, show that $f(s,t) = \phi(s) + \psi(t)$ for some $\phi, \psi : \mathbb{R} \rightarrow \mathbb{R}$. [Hint: consider $f(s,t) - f(s,c)$ as a function of s.]
   (iii) If $(D_{22}f)(z) = 0$ for all $z \in A$, show that $f(s,t) = \phi(s) + \psi(s)t$ for suitable ϕ and ψ.

3. Let $f : \mathbb{R}^2 \rightarrow \mathbb{R}$ be differentiable. Define $g : \mathbb{R} \rightarrow \mathbb{R}$ by $g(s) = f(s, c - s)$, where c is a constant. Write down an expression for $g'(s)$. If $D_1f = D_2f$ on $\mathbb{R}^2$, deduce that $f(s,t) = h(s+t)$ for some $h : \mathbb{R} \rightarrow \mathbb{R}$.

4. A function $f : \mathbb{R}^2 \rightarrow \mathbb{R}$ is said to be harmonic if all its second partial derivatives are continuous and $D_{11}f + D_{22}f = 0$.
   (i) What functions of the form $f(s,t) = as^2 + 2bst + ct^2$ are harmonic?
   (ii) What can be deduced about the points z at which a harmonic function f satisfies $(D_1f)(z) = (D_2f)(z) = 0$?
   (iii) If $U = \{(s,t) : s^2 + t^2 \leq 1\}$, what can be said about the points where $\sup f(U)$ and $\inf f(U)$ are attained?

5. Let $f : \mathbb{R}^2 \rightarrow \mathbb{R}$ be given by
$$f(s,t) = \begin{cases} [st(s^2 - t^2)]/[s^2 + t^2] & (s,t) \neq (0,0) \\ 0 & s = t = 0. \end{cases}$$

Show that the second partial derivatives $D_{12}f$ and $D_{21}f$ both exist at $(0,0)$, but that they are not equal there.

6. Find the critical points (i.e., the points $z \in \mathbb{R}^2$ at which both $(D_1 f)(z)$ and $(D_2 f)(z)$ are zero) of the following functions, and determine their nature:
   (i) $f(s,t) = s^2 + 4st$
   (ii) $f(s,t) = s^2 + 4st + 2t^2 - 2t$
   (iii) $f(s,t) = (s-1)^4 + (s-t)^4$.

7. Show that the function $f : \mathbb{R}^2 \to \mathbb{R}$ given by $f(s,t) = 2s + 4t - s^2$ has a point z for which $(D_1 f)(z) = (D_2 f)(z) = 0$. Show that this function has no relative maxima or minima.

8. Show that the following functions have $D_1 f$ and $D_2 f$ zero at the point $(0,0)$;
   (i) $f(s,t) = s^3 t^3$
   (ii) $f(s,t) = s^4 t^4$.
   Show (in the notation of **9.12**) that $\Delta = 0$ in both cases. Show also that $(0,0)$ is a saddle point in (i) and a relative minimum in (ii). This example shows that $\Delta = 0$ really does convey no information about the behaviour of the function at the relevant point.

9. Use the ideas of **9.11** to rewrite part(ii) of **9.10** purely in terms of the derivative operator. Try to prove a general version of Taylor's theorem (we only handled the case for $n = 1$ if you are familiar with the notation for the one-dimensional version).

10

Differentiation – a more abstract viewpoint

This chapter discusses the possibility of differentiating a mapping which carries elements in one normed linear space X into another, Y. As such, it contains some of the results of chapter **9** as special cases, since in that chapter we discussed the case $X = (\mathbb{R}^2, \|\cdot\|_\infty)$ while Y was $(\mathbb{R}, |\cdot|)$. Our major source of applications in this chapter occurs when $X = (\mathbb{R}^p, \|\cdot\|_\infty)$ and $Y = (\mathbb{R}^q, \|\cdot\|_\infty)$, where $p, q > 1$. Again the idea behind the notion of a derivative is the desire to approximate a function locally by a linear function. This is done in an analogous manner to the previous chapter.

Definition 10.1 *Let X and Y be normed linear spaces. A mapping $f : X \to Y$ is said to be differentiable at x_0 in X if there is a continuous linear mapping $T : X \to Y$ such that, given $\epsilon > 0$ there is a $\delta > 0$ so that*

$$\|f(x_0 + h) - f(x_0) - Th\| \le \epsilon \|h\| \quad \text{for all } h \in X, \|h\| \le \delta.$$

The mapping T is called the derivative of f at x_0 and is written $Df(x_0)$.

Note that $Df(x_0)$ is a linear operator in $\mathcal{B}(X, Y)$, and that this operator, when it exists, is unique. As such, the theorems of chapter **8** are applicable. Also the expression $Df(x_0)(z)$ will be used to denote the action of the linear operator $Df(x_0)$ on the element z in X. The norm of $Df(x)$ is defined in the usual way:

$$\|Df(x)\| = \sup_{\|z\|=1} \|Df(x)(z)\|.$$

Notice that **8.5** implies that $Df(x)$, if it exists, is always bounded. It is important that this definition of a derivative should correspond to the usual one, when the mapping f is from $\mathbb{R}$ to $\mathbb{R}$ or from $\mathbb{R}^2$ to $\mathbb{R}$. By refering back to chapter **9**, we see that if $f : \mathbb{R}^2 \to \mathbb{R}$ has a gradient vector (A, B) at some point $x \in \mathbb{R}^2$, then the operator $Df(x) : \mathbb{R}^2 \to \mathbb{R}$ would be defined by

$$Df(x)(h, k) = Ah + Bk, \qquad (h, k) \in \mathbb{R}^2.$$

The discussion in that chapter of the simple univariate derivative shows that if $f : \mathbb{R} \to \mathbb{R}$ has $f'(x_0) = \alpha$, then the appropriate choice for $Df(x_0)$ is the mapping defined by $Df(x_0)(h) = \alpha h$, $h \in \mathbb{R}$.

Consider the simplest example which is not contained in the discussions of the previous chapter. Thus take $f : \mathbb{R}^2 \to \mathbb{R}^2$. Then f can be written as $f(x) = [f_1(x), f_2(x)]$ where $x \in \mathbb{R}^2$ and f_1, f_2 are mappings of $\mathbb{R}^2$ into $\mathbb{R}$. The operator $Df(x)$ is required to be a linear mapping from $\mathbb{R}^2$ into $\mathbb{R}^2$, and so is a 2×2 matrix. Write

$$Df(x_0) = \begin{pmatrix} t_{11} & t_{12} \\ t_{21} & t_{22} \end{pmatrix} \quad \text{and} \quad h = (h_1, h_2), \ h \in \mathbb{R}^2.$$

Then the inequality in **10.1** may be written as

$$\left\| \begin{pmatrix} f_1(x_0 + h) \\ f_2(x_0 + h) \end{pmatrix} - \begin{pmatrix} f_1(x_0) \\ f_2(x_0) \end{pmatrix} - \begin{pmatrix} t_{11}h_1 + t_{12}h_2 \\ t_{21}h_1 + t_{22}h_2 \end{pmatrix} \right\| \le \epsilon \|h\|.$$

If we employ the maximum norm then the above inequality can only be satisfied if

$$|f_1(x_0 + h) - f_1(x_0) - (t_{11}h_1 + t_{12}h_2)| \le \epsilon \|h\|,$$

and

$$|f_2(x_0 + h) - f_2(x_0) - (t_{21}h_1 + t_{22}h_2)| \le \epsilon \|h\|.$$

Thus given $\epsilon > 0$ there is a $\delta > 0$ such that the above *pair* of inequalities hold for all $h \in \mathbb{R}^2$ with $\|h\| \le \delta$. But this is precisely the statement that f_1 and f_2 are differentiable at x_0 with gradient vectors (t_{11}, t_{12}) and (t_{21}, t_{22}) respectively. Now **9.4** tells us that $t_{11} = (D_1 f_1)(x_0)$, $t_{12} = (D_2 f_1)(x_0)$, $t_{21} = (D_1 f_2)(x_0)$ and $t_{22} = (D_2 f_2)(x_0)$. Thus

$$Df(x_0) = \begin{pmatrix} (D_1 f_1)(x_0) & (D_2 f_1)(x_0) \\ (D_1 f_2)(x_0) & (D_2 f_2)(x_0) \end{pmatrix}.$$

This matrix is often called the Jacobian matrix. Its determinant is called simply the Jacobian. Straightforward generalisations of the above argument and those of chapter **9** show that if $f : \mathbb{R}^p \to \mathbb{R}^q$, then the derivative of f at x_0 (if it exists) is given by the Jacobian matrix $Df(x_0)$ whose ij^{th} element is

$$D_j f_i(x_0) \quad \text{where} \quad f = (f_1, f_2, \ldots, f_p) \quad \text{and} \quad 1 \le i \le q, \ 1 \le j \le p.$$

The next few results explore some of the properties of differentiable functions. Their proofs are often strongly reminiscent of those in the previous chapter.

Lemma 10.2 *If $f : X \to Y$ is differentiable at $x_0 \in X$, then f is continuous at x_0.*

Proof. Suppose the derivative of f at x_0 is $T = Df(x_0)$. Take $\epsilon > 0$ and suppose $\delta > 0$ is chosen so that

$$\|f(x_0 + h) - f(x_0) - Th\| < \|h\|$$

for all $h \in X$ such that $\|h\| < \delta$. Take

$$\delta' < \min(\delta, \frac{\epsilon}{1 + \|T\|}).$$

Then for all $h \in X$ with $\|h\| < \delta'$ we have

$$
\begin{aligned}
\|f(x_0 + h) - f(x_0)\| &\le \|f(x_0 + h) - f(x_0) - Th\| + \|Th\| \\
&\le \|h\| + \|T\|\|h\| \\
&= (1 + \|T\|)\|h\| \\
&\le (1 + \|T\|)\delta' \\
&< \epsilon.
\end{aligned}
$$

Hence f is continuous at x_0. ∎

The following exercises consist entirely of elementary facts about the derivative. Hints as to their solutions can be gleaned from chapter **9**.

Exercises

1. If $f : X \to Y$ is a continuous linear mapping, then it is differentiable at every point in X and $Df(x) = f(x)$ for all $x \in X$.

2. Suppose $f : X \to Y$ is differentiable at x_0 and $Df(x_0) = T$. Show that for any $h \in X$,

$$Th = \lim_{\lambda \to 0} \frac{1}{\lambda} [f(x_0 + \lambda h) - f(x_0)].$$

3. Suppose that $f : X \to \mathbb{R}$ is a differentiable function which has a local maximum or minimum at $x_0 \in X$. Show that $Df(x_0) = 0$.

4. Suppose that $\phi : X \to \mathbb{R}$ is a linear mapping. Suppose also that $f : X \to \mathbb{R}$ is differentiable at $x_0 \in X$ and that

$$\phi(x - x_0) \le f(x) - f(x_0) \quad \text{for all } x \in X.$$

Show that $Df(x_0) = \phi$.

5. Let $f, g : X \to \mathbb{R}$ have derivatives A and B respectively at $x_0 \in X$. Obtain a reasonable expression for the derivative of $f.g$ at x_0 and establish that your formula is correct.

Suppose $f : X \to Y$ is differentiable at $x_0 \in X$ with derivative S. By **10.1**, this means roughly that for small h (i.e., h close to θ_X in X)

$$f(x_0 + h) \cong f(x_0) + Sh.$$

If $g : Y \to Z$ is differentiable at $y_0 = f(x_0)$ in Y with derivative T, then for small k

$$g(y_0 + k) \cong g(y_0) + Tk,$$

and so we can write

$$
\begin{aligned}
g[f(x_0 + h)] \cong g[f(x_0) + Sh] &= g(y_0 + Sh) \\
&\cong g(y_0) + (T \circ S)h \\
&= g[f(x_0)] + (T \circ S)h,
\end{aligned}
$$

that is,

$$(g \circ f)(x_0 + h) - (g \circ f)(x_0) \cong (T \circ S)h.$$

If $g \circ f$ is differentiable at x_0, then its derivative operator R should approximate $g \circ f$ well in a neighbourhood of x_0. The above equation suggests that $R = T \circ S$. This is in fact the case, and the result is often called the *chain rule* for differentiation. In its statement, the customary compression of notation for linear operators is made, so that $T \circ S$ is simply written TS.

Theorem 10.3 *Let X, Y and Z be normed linear spaces. Let f mapping X to Y be differentiable at $x_0 \in X$ with derivative S, and let g mapping Y to Z be differentiable at $f(x_0) \in Y$ with derivative T. Then $g \circ f$ is differentiable at x_0 with derivative TS.*

Proof. Set $y_0 = f(x_0)$ and define $\phi : X \to Y$, $\psi : Y \to Z$ by

$$\phi(h) = f(x_0 + h) - f(x_0) - Sh, \quad (h \in X),$$

and

$$\psi(k) = g(y_0 + k) - g(y_0) - Tk, \quad (k \in Y).$$

Then

$$
\begin{aligned}
g[f(x_0 + h)] - g(y_0) &= g[f(x_0) + Sh + \phi(h)] - g(y_0) \\
&= g[y_0 + Sh + \phi(h)] - g(y_0) \\
&= T[Sh + \phi(h)] + \psi[Sh + \phi(h)].
\end{aligned}
$$

Hence

$$g[f(x_0 + h)] - g[f(x_0)] - TSh = T\phi(h) + \psi[Sh + \phi(h)] = \theta(h),$$

say. Given $\epsilon > 0$, choose ϵ' such that $(1 + \|S\| + \|T\|)\epsilon' \le \epsilon$ and $0 < \epsilon' < 1$. To show that $g \circ f$ is differentiable at x_0 with derivative TS, **10.1** demands that we show that $\|\theta(h)\| \le \epsilon \|h\|$ for sufficiently small $\|h\|$. Since f is differentiable at x_0 and g is differentiable at y_0, there is a $\delta > 0$ such that

$$\|\phi(h)\| \le \epsilon' \|h\| \quad \text{and} \quad \|\psi(k)\| \le \epsilon' \|k\|,$$

for all $\|h\| \le \delta$ and $\|k\| \le \delta$. Now further restrict h so that

$$\|h\| \le \frac{\delta}{\|S\| + 1}.$$

Then, for such h,

$$
\begin{aligned}
\|Sh + \phi(h)\| &\le \|Sh\| + \|\phi(h)\| \\
&\le \|S\|\|h\| + \|\phi(h)\| \\
&\le \|S\|\|h\| + \epsilon'\|h\| \\
&\le (\|S\| + 1)\|h\| \\
&\le \delta.
\end{aligned}
$$

Since $\|Sh + \phi(h)\| \le \delta$ we have

$$\begin{aligned}
\|\psi[Sh + \phi(h)]\| &\le \epsilon'\|Sh + \phi(h)\| \\
&\le \epsilon'(\|Sh\| + \|\phi(h)\|) \\
&\le \epsilon'(\|S\|\|h\| + \epsilon'\|h\|) \\
&\le \epsilon'(\|S\| + 1)\|h\|.
\end{aligned}$$

Using these facts in the definition of $\theta(h)$, we obtain, for all $\|h\| \le \delta/(\|S\| + 1)$,

$$\begin{aligned}
\|\theta(h)\| &\le \|T\phi(h)\| + \|\psi[Sh + \phi(h)]\| \\
&\le \|T\|\|\phi(h)\| + \epsilon'(\|S\| + 1)\|h\| \\
&\le \epsilon'\|T\|\|h\| + \epsilon'(\|S\| + 1)\|h\| \\
&= \epsilon'(1 + \|S\| + \|T\|)\|h\| \\
&\le \epsilon\|h\|.
\end{aligned}$$

Thus $g \circ f$ is differentiable at x_0, with derivative TS. ∎

It is often helpful to reinterpret this result in an alternative notation. Recall, firstly, that the derivative of $g \circ f$ should be a member of $\mathcal{B}(X, Z)$. Theorem **10.3** says that

$$D(g \circ f)(x_0) = TS = Dg(f(x_0))Df(x_0).$$

Alternatively,

$$D(g \circ f) = (Dg \circ f)Df.$$

Our notation is becoming somewhat strained here, since the convention we have adopted is to write $Df(x_0)$ for the derivative of the function f at x_0. However, the composition $D(g \circ f)(x_0)$ would be somewhat ambiguous unless $g \circ f$ is enclosed in parentheses.

Now let $X = Y = Z = \mathbb{R}^p$. If

$$J_f(x_0) = \det\left(Df(x_0)\right),$$

then **10.3** induces a relationship on the Jacobians. This is

$$J_{g \circ f}(x_0) = J_g(f(x_0))J_f(x_0).$$

If, for fixed a in $\mathbb{R}^q$, $J_g(a)$ is regarded as a mapping from $\mathbb{R}^q$ to $\mathbb{R}^s$, then the above formula can be written again as a composition:

$$J_{g \circ f}(x_0) = (J_g \circ f)(x_0)J_f(x_0).$$

Corollary 10.4 *Let f_1 and f_2 be mappings of $\mathbb{R}$ into $\mathbb{R}$ which are differentiable at $t_0 \in \mathbb{R}$. Let $f_i(t_0) = x_i$, $i = 1, 2$. Suppose $g : \mathbb{R}^2 \to \mathbb{R}$ is differentiable at (x_1, x_2) and that $G : \mathbb{R} \to \mathbb{R}$ is defined by $G(t) = g[f_1(t), f_2(t)]$, $t \in \mathbb{R}$. Then*

$$G'(t_0) = (D_1 g)(x_1, x_2) f_1'(t_0) + (D_2 g)(x_1, x_2) f_2'(t_0).$$

Proof. This result has only to be interpreted in the light of **10.3**. To do this, we have to understand the function $G : \mathbb{R} \to \mathbb{R}$ as the composition of two mappings. Define a mapping $H : \mathbb{R} \to \mathbb{R}^2$ by $H(t) = [f_1(t), f_2(t)]$. Then $G = g \circ H$. Suppose g has derivative T at (x_1, x_2). Then T is given by the (2×1) matrix $((D_1 g)(x_1, x_2), (D_2 g)(x_1, x_2))$. If H has derivative S at t_0 then S is given by the (1×2) matrix

$$\begin{pmatrix} f_1'(t_0) \\ f_2'(t_0) \end{pmatrix}.$$

Now **10.3** says that for $G : \mathbb{R} \to \mathbb{R}$,

$$G'(t_0) = TS = (D_1 g)(x_1, x_2) f_1'(t_0) + (D_2 g)(x_1, x_2) f_2'(t_0)$$

as required. ∎

This result is often expressed in somewhat different notation. To understand this, write $g = g(f_1, f_2)$, when **10.4** could be written

$$\frac{dg}{dt} = \frac{\partial g}{\partial f_1} \frac{df_1}{dt} + \frac{\partial g}{\partial f_2} \frac{df_2}{dt}.$$

Although this is the customary way of expressing **10.4** in elementary texts on calculus, the notation really is very poor. For example, the notation 'g' is trying to serve two roles. It stands for the function from $\mathbb{R}^2 \to \mathbb{R}$ as it did in **10.4**, but it is at the same time representing the univariate function labelled 'G' in **10.4**. Additionally, f_1 and f_2 are playing two different roles (as the arguments of the function g, and as mappings from $\mathbb{R}$ to $\mathbb{R}$).

Example Suppose $G : \mathbb{R} \to \mathbb{R}$ is defined by $G(t) = g(\cos t, \sin t)$ where g is any differentiable function from $\mathbb{R}^2$ into $\mathbb{R}$. Then

$$G'(t) = -\sin t (D_1 g)(\cos t, \sin t) + \cos t (D_2 g)(\cos t, \sin t).$$

Thus if $g : \mathbb{R}^2 \to \mathbb{R}$ is defined by $g(a, b) = a^2 + b^2$, then

$$G'(t) = -\sin t.2 \cos t + \cos t.2 \sin t = 0.$$

This can be checked easily by direct substitution, since

$$G(t) = \sin^2 t + \cos^2 t = 1.$$

This chapter now concludes with some theorems on the differentiation of inverse functions. Consider for a moment the univariate case. Let $\mathbb{R}^+ = \{x \in \mathbb{R} : x > 0\}$. The functions $f : \mathbb{R}^+ \to \mathbb{R}^+$ and $g : \mathbb{R}^+ \to \mathbb{R}^+$ given by $f(x) = e^x$ and $g(x) = \ln x$ have the property that

$$(f \circ g)(x) = (g \circ f)(x) = x,$$

that is, they are each the inverse function of the other. Both these functions are differentiable everywhere on their domains of definition. Furthermore, $f'(1) = e$ while $g'(f(1)) = 1/e = 1/f'(1)$. This is not a coincidence, but is the realisation of a natural relationship between the derivatives of the two functions. This relationship follows from the univariate chain rule, since differentiating both sides of the equality $g[f(x)] = x$ with respect to x gives

$$g'[f(x)]f'(x) = 1.$$

that is

$$g'[f(x)] = \frac{1}{f'(x)}.$$

Let $f : X \to Y$ and $g : Y \to X$ be functions which are differentiable at $x_0 \in X$ and $f(x_0) \in Y$ respectively. If $g \circ f$ is the identity map on X, then **10.3** tells us that (writing I_X for the identity map on X)

$$I_X = D(g \circ f)(x_0) = Dg[f(x_0)] \circ Df(x_0).$$

This suggests that $Dg[f(x_0)] = [Df(x_0)]^{-1}$, or alternatively, that $Dg(y_0) = [Df(g(y_0))]^{-1}$. In order to draw this conclusion, we made assumptions not only about f, but also about its inverse g. The aim of the next few results is to make assumptions only on the function f, and to try to deduce the differentiablity of g from there. A secondary aim is to produce a theorem of sufficient generality to act as an essential tool in the proof of the implicit function theorem, which is coming in chapter **12**. It will help in obtaining concise statements of the following results if we introduce the following notation.

Definition 10.5 *Let W be an open subset of a normed linear space X. The mapping $f : W \to Y$, where Y is a normed linear space, is said to belong to the class $C^1(W)$ if $Df(x)$ exists for all $x \in W$, and the mapping $x \mapsto Df(x)$ is continuous on W.*

Recall that for each x, $Df(x)$, if it exists, is a continuous linear mapping from X to Y. Thus the mapping $x \mapsto Df(x)$ associates each x with a continuous, linear mapping of X into Y, that is, with a member of $\mathcal{B}(X, Y)$. It follows that the continuity of the mapping $x \mapsto Df(x)$ is defined by using the given norm on X, and the operator norm in $\mathcal{B}(X, Y)$, viz:

$$\|Df(x)\| = \sup_{\|z\|=1} \|Df(x)(z)\|.$$

Notice also that the mapping $x \mapsto Df(x)$ is *not* in general a linear mapping (nor is it necessarily continuous). It is only the range of this mapping that contains continuous linear mappings. We begin with an elementary result.

Theorem 10.6 *Let A and B be open subsets of normed linear spaces X and Y respectively. Let f be a bijection of A onto B, such that the inverse function g is continuous. If f is differentiable at a in A and the mapping $Df(a)$ has a continuous inverse, then g is differentiable at $f(a)$ and $Dg(f(a)) = [Df(a)]^{-1}$.*

Proof. Set $T = Df(a)$, and take $\epsilon_0 > 0$. Since T^{-1} is continuous in $\mathcal{B}(Y, X)$ it follows from **8.5** that T^{-1} is bounded. Choose $\epsilon > 0$ so that

$$\epsilon \leq \min\left\{\frac{1}{2\|T^{-1}\|}, \frac{\epsilon_0}{2\|T^{-1}\|^2}\right\}.$$

Since f is differentiable at a, there is a $\delta > 0$ such that, for $\|h\| < \delta$,

$$\|f(a+h) - f(a) - Th\| \leq \epsilon\|h\|. \tag{$\dagger$}$$

Set $b = f(a)$ and choose $k \in Y$ such that $b + k \in B$. Write $g(b+k) = a + h$. Since g is continuous, there is a $\delta' > 0$ such that if $\|k\| \leq \delta'$ then $\|g(b+k) - g(b)\| \leq \delta$, i.e. $\|h\| \leq \delta$. By ($\dagger$),

$$\|f[g(b+k)] - f[g(b)] - Th\| \leq \epsilon\|h\| \quad \text{for} \quad \|k\| \leq \delta'.$$

That is

$$\|b + k - b - Th\| \leq \epsilon\|h\|,$$

so that

$$\|k - Th\| \leq \epsilon \|h\|.$$

Hence

$$
\begin{aligned}
\|h\| - \|T^{-1}k\| &\leq \|T^{-1}k - h\| \\
&= \|T^{-1}(k - Th)\| \\
&\leq \|T^{-1}\|\|k - Th\| \\
&\leq \|T^{-1}\|\epsilon\|h\| \\
&\leq \|h\|/2.
\end{aligned}
$$

This gives $\|h\| \leq 2\|T^{-1}k\|$. Now use the fact that $h = g(b+k) - g(b)$ to write

$$
\begin{aligned}
\|T^{-1}k - g(b+k) - g(b)\| &= \|T^{-1}k - h\| \\
&\leq \epsilon\|T^{-1}\|\|h\| \\
&\leq 2\epsilon\|T^{-1}\|\|T^{-1}k\| \\
&\leq 2\epsilon\|T^{-1}\|^2\|k\| \\
&\leq \epsilon_0\|k\|.
\end{aligned}
$$

Hence g is differentiable at $f(a)$ with derivative T^{-1} as required. ∎

The next result is quite deep. It is needed for the proof of the implicit function theorem in chapter **12**.

Theorem 10.7 *Let W be an open subset of a Banach space X, and suppose $f : W \rightarrow X$ is in $C^1(W)$. Let a be a point in W such that $Df(a)$ is a bijection. Then there exists an open set U containing a such that $V = f(U)$ is open set containing $f(a)$, and the restriction of f to U is a bijection onto V with inverse g in $C^1(V)$. Furthermore,*

$$Dg(v) = [Df\{g(v)\}]^{-1} \quad \text{for all } v \text{ in } V.$$

Proof. We begin by treating a special case. Assume $a = \theta$, $f(\theta) = \theta$ and $Df(\theta) = I$. The overall strategy is as follows. We begin by making an initial guess for U. This enables us to construct a suitable V. From this V our initial guess is refined. Start by fixing α, where $0 < \alpha < 1$. Define $\phi : W \rightarrow X$ by

$$\phi(x) = x - f(x), \qquad (x \in W).$$

Then $\phi \in C^1(W)$ and $D\phi(\theta) = \theta$. This last statement means that $D\phi(\theta)$ is the zero operator in $\mathcal{B}(X, X)$. Since the mapping $x \mapsto D\phi(x)$ is continuous, there is an open ball $B \subset W$ and centred on θ such that

$$\|D\phi(x)\| < \alpha \quad \text{for all } x \text{ in } B.$$

The next stage in the proof is to identify a suitable choice for U and V. We begin by determining a subset of X on which f is invertible. Fix two points x and y in B, and set

$$x_\lambda = (1 - \lambda)x + \lambda y, \qquad (\lambda \in \mathbb{R}).$$

Define $\psi : [0, 1] \to X$ by $\psi(\lambda) = \phi(x_\lambda)$, $\lambda \in [0, 1]$. Then by the chain rule (see **10.3**),

$$D\psi(\lambda) = D\phi(x_\lambda) \circ (y - x),$$

that is, $D\psi(\lambda)$ is the linear mapping from $[0, 1]$ to X defined by

$$D\psi(\lambda)(s) = [D\phi(x_\lambda) \circ (y - x)](s) = D\phi(x_\lambda)(y - x), \quad s \in [0, 1].$$

Hence

$$\begin{aligned}
\|D\psi(\lambda)(s)\| &= \|D\phi(x_\lambda)(y - x)\| \\
&\leq \|D\phi(x_\lambda)\|\|y - x\| \\
&< \alpha\|y - x\|.
\end{aligned}$$

It now follows from the mean value theorem (see exercise 2) that for some $\lambda \in (0, 1)$,

$$\|\phi(y) - \phi(x)\| = \|\psi(1) - \psi(0)\| \leq \|D\psi(\lambda)\|\|1 - 0\| < \alpha\|y - x\|.$$

Hence,

$$\begin{aligned}
\|f(y) - f(x)\| &= \|y - \phi(y) - x + \phi(x)\| \\
&\geq \|x - y\| - \|\phi(y) - \phi(x)\| \\
&\geq (1 - \alpha)\|x - y\|, \qquad\qquad (\ddagger)
\end{aligned}$$

for all $x, y \in B$. This has the immediate consequence that f is injective on B. Also, if $g : f(B) \to B$ is defined by $g[f(x)] = x$ for $x \in B$, then $(\ddagger)$ may be written as

$$(1 - \alpha)\|g[f(x)] - g[f(y)]\| \leq \|f(x) - f(y)\|.$$

Since $0 < \alpha < 1$ this statement implies that g is a Lipschitz mapping, and so is certainly continuous.

We have now constructed the set B such that $f : B \to f(B)$ is a bijection. It must be established that $g = f^{-1}$ is continuously differentiable on an appropriate set. In order that f be a bijection, we had to restrict our attention to an open ball containing the point θ. We now turn to the point $f(\theta) \in f(B)$. (Recall that we are still assuming $f(\theta) = \theta$.) In trying to ensure g is continuously differentiable we shall in fact have to take a smaller set than $f(B)$.

We claim $f(B) \supset (1 - \alpha)B = \{(1 - \alpha)x : x \in B\}$. To see this, take $y \in (1 - \alpha)B$. A sequence $\{x_n\}$ is going to be defined inductively. Put $x_0 = \theta$, $x_1 = y$. Suppose $n \geq 1$ and $x_0, \ldots, x_n$ have been constructed so that

$$x_i = y + \phi(x_{i-1}) \quad \text{and} \quad \|x_i - x_{i-1}\| \leq \alpha^{i-1}\|y\| \quad (1 \leq i \leq n). \qquad (*)$$

Then

$$
\begin{aligned}
\|x_n\| &= \left\| \sum_{i=1}^{n}(x_i - x_{i-1}) \right\| \\
&\leq \sum_{i=1}^{n} \|x_i - x_{i-1}\| \\
&\leq \sum_{i=1}^{n} \alpha^{i-1}\|y\| \\
&\leq (1 - \alpha)^{-1}\|y\|. \qquad (**)
\end{aligned}
$$

This implies that, since $y = (1 - \alpha)z$ for some $z \in B$, $\|x_n\| \leq \|z\|$ and so $x_n \in B$. Thus $\phi(x_n)$ is well-defined and we can take

$$x_{n+1} = y + \phi(x_n).$$

It follows that

$$\|x_{n+1} - x_n\| = \|\phi(x_n) - \phi(x_{n-1})\| < \alpha\|x_n - x_{n-1}\| \leq \alpha^n\|y\|.$$

Now our conditions certainly hold for the pair x_0, x_1 and the above argument shows that the sequence $\{x_n\}$ can be defined recursively so that $x_0 = \theta$, $x_1 = y$ and the conditions $(*)$ hold for all $i \geq 1$. This sequence is Cauchy, since for $n > m \geq 0$,

$$\|x_n - x_m\| = \left\| \sum_{i=m+1}^{n}(x_i - x_{i-1}) \right\|$$

$$\leq \sum_{i=m+1}^{n} \|x_i - x_{i-1}\|$$

$$\leq \sum_{i=m+1}^{n} \alpha^{i-1}\|y\|$$

$$\leq \alpha^m \|y\| \sum_{i=1}^{n-m-1} \alpha^{i-1}$$

$$\leq \frac{\alpha^m}{1-\alpha}\|y\|.$$

Hence $\{x_n\}$ converges to a point p in X. But

$$\|p\| = \lim_{n\to\infty} \|x_n\| \leq (1-\alpha)^{-1}\|y\|$$

by (**), and so $p \in B$. Now by the definition, ϕ is a continuous function on W, and so by **2.10**

$$p - f(p) = \phi(p) = \lim_{n\to\infty} \phi(x_n) = \lim_{n\to\infty}(x_{n+1} - y) = p - y.$$

Hence $y = f(p)$ and so $y \in f(B)$ as claimed.

Take $V = (1-\alpha)B$. Since we have just established $V \subset f(B)$, and our earlier reasoning showed f is injective on B, we take $U = g(V) = B \cap f^{-1}(V)$. Since f is continuous, U, being the intersection of the two open sets, B and $f^{-1}(V)$, is open. Since $f(\theta) = \theta$ and $\theta \in V$, U contains θ.

It remains to show g is continuously differentiable on V. Since $\alpha < 1$,

$$\|I - Df(z)\| < 1$$

for all $z \in B$. Consequently, by **8.13**, $Df(z)$ has a continuous inverse for each z in U. We can therefore apply **10.6** with $A = U$, $B = V$ to conclude that $Dg(y) = [Df\{g(y)\}]^{-1}$, for all $y \in V$. Now $Dg(y)$ is the composition of the three mappings $g : V \to U$, $x \mapsto Df(x)$ from U to $\mathcal{B}(X)$ and the inversion map. The first of these maps has been established as continuous in the course of the proof, the second is continuous by hypothesis, and the continuity of the third follows from **8.14**. The continuity of $Dg(y)$ now follows from **2.9**.

Finally, there remains the case when the assumptions $a = \theta$, $f(\theta) = \theta$ and $Df(\theta) = I$ are not valid. Set $T = Df(a)$, $W_1 = \{w - a : w \in W\}$ and

$$f_1(x) = T^{-1}[f(a + x) - f(a)], \qquad (x \in W_1).$$

Then $\theta \in W_1$, $f_1(\theta) = \theta$ and $Df_1(\theta) = I$. Thus by our previous arguments, there exist open sets U and V such that the conclusions of the theorem hold for f_1 and its inverse g_1 say. Now writing

$$f(a + x) = Tf_1(x) + f(a), \qquad (x \in W_1),$$

enables us to see that the desired conclusions hold on f. ∎

Exercises

1. Let X and Y be Banach spaces, and define $P : X \times Y \to X$ by $P(x,y) = x$ for all $x \in X$. Show that $P \in C^1(X \times Y)$.

2. Let a, b belong to the normed linear space X. Let $f : X \to Y$ be a mapping from X to the normed linear space Y such that f is differentiable at every point on the line segment joining a to b. Show that there is a point c on this same line segment such that

$$f(b) - f(a) = Df(c)(b - a).$$

[Hint: Look back to the mean value theorem in the previous chapter.]

3. Let X and Y be Banach spaces and let $f : X \times Y \to Y$ be in $C^1(W)$ where W is an open set in $X \times Y$. Define $h : X \times Y \to X \times Y$ by $h(x,y) = (x, f(x,y))$ for all $(x,y) \in X \times Y$. Show that $h \in C^1(W)$ and that $Dh(x,y) = (x, Df(x,y))$ for all $(x,y) \in W$.

11

The Riemann integral

We begin with a brief discussion of integration of univariate functions. In this setting, integration is bound up with the idea of calculating area. The formulae for calculating the area of simple plane figures, such as rectangles, triangles, parallelograms, are well-known. What Riemann integration does is to first of all extend the notion of area to more general plane figures, and then give methods for calculating such areas. For example, consider the shaded area A as indicated in 11.1. There the three sides of the shape are linear, while the fourth side is defined by some continuous function f. We want to calculate the area A. There are a variety of ingenious physical solutions to this problem. For example, one could construct a vessel whose base was the area A, and which had sides perpendicular to A. If water is then poured into such a vessel to a depth of say 1 centimetre, then the volume of the water needed in cubic centimetres would be the same as the area A in square centimetres. Such a process has two drawbacks. Firstly, it is laborious, involving the physical construction of the vessel. Secondly, it will always be inaccurate, even when the fourth side is defined by a simple function. For example, if the function f is linear, then the shape of A is a trapezium, whose area can be calculated exactly by other means. The vessel construction will only allow us to estimate this area, the precision depending on several factors (the accuracy of construction of the vessel, allowance for the surface tension of the water producing a meniscus, and so on). The approach to computation of area via the Riemann integral is superior to such methods, since all the simple plane figures (rectangles, triangles, etc.) will have their areas computed exactly. In addition, many hitherto unknown areas

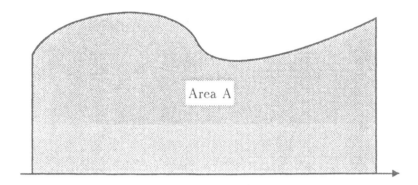

Figure 11.1: A typical plane area.

can be computed exactly by the Riemann integral. Finally, most areas can be estimated to a high accuracy using the ideas of Riemann integration and the aid of a digital computer.

The idea is very simple. Suppose we want to calculate the area A in 11.1. Then we first of all define a *partition* of the interval $[a, b]$. This is a set of points $s_0, s_1, \ldots, s_n$ where

$$a = s_0 < s_1 < \ldots < s_{n-1} < s_n = b.$$

Suppose that f is a positive, bounded, real-valued function on $[a, b]$. Then the *lower sum* is defined as

$$\sum_{i=1}^{n} \inf\{f(s) : s \in [s_{i-1}, s_i]\}(s_i - s_{i-1}).$$

This defines the area shown in 11.2 where for convenience n is assumed to be 9. In a similar manner we define the *upper sum* as

$$\sum_{i=1}^{n} \sup\{f(s) : s \in [s_{i-1}, s_i]\}(s_i - s_{i-1}).$$

Again this area is illustrated in 11.3 with the same value of n. Now it is clear that whatever partition we take in $[a, b]$, the value of the area A will always lie between the lower sum and the upper sum corresponding to the given partition. Also, intuitively, as the number of points in the partition grows, the upper and lower sums ought to establish a 'pincer

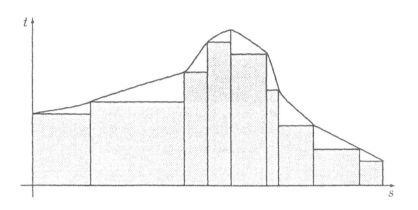

Figure 11.2: A lower sum.

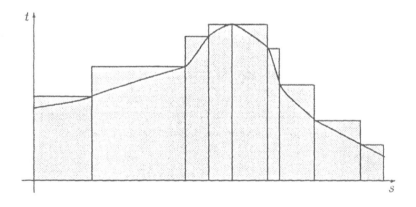

Figure 11.3: An upper sum.

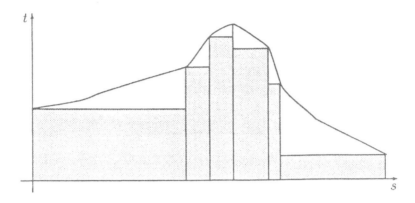

Figure 11.4: A bad distribution of partition points.

grip' on the value of A. A moment's reflection, however, shows that we must achieve some control on the distribution of points. It is easy to construct partitions which estimate the required area very poorly, and Figure 11.4 illustrates just such a partition. The description of how the partition must reflect the variation of the function f is a difficult task, but fortunately there is a neat way to avoid this problem, in theory at least. In practice the linking of the partition to the behaviour of the function is the basis of adaptive numerical integration routines, but we will not consider such matters here. Instead we define the *lower integral* of the function f to be the supremum of the lower sums over *all* possible partitions of the interval $[a, b]$. Since each lower sum is never greater than the value of A, the lower integral is also less than this value (and exists). The *upper integral* is defined to be the infimum of the upper sums over all possible partitions of the interval $[a, b]$. The upper integral is greater than or equal to the value of the area A, so that

$$\text{lower integral} \leq \text{area } A \leq \text{upper integral.}$$

We would expect in many cases to find that the lower and upper integrals have the same value. In this case, we pronounce the function f to be *Riemann-integrable* on $[a, b]$, and declare the area A to be this value.

Now we extend this programme to functions $f : \mathbb{R}^2 \to \mathbb{R}$. In this case, we can think of f as defining a surface in $\mathbb{R}^3$. When f is non-negative, the process of Riemann integration defines and computes a volume, although we will not restrict our considerations to non-negative functions. Throughout the following discussions $[a, b]$ and $[c, d]$ will be real intervals and $f : \mathbb{R}^2 \to \mathbb{R}$ will be a bounded function on $[a, b] \times [c, d]$. Suppose partitions P_1 of $[a, b]$ and P_2 of $[c, d]$ have been chosen, where

$$P_1 \text{ is the partition} \quad a = s_0 < s_1 < \ldots < s_n = b;$$
$$P_2 \text{ is the partition} \quad c = t_0 < t_1 < \ldots < t_n = d.$$

Write

$$A_i = [s_{i-1}, s_i], \quad l_i = s_i - s_{i-1}, \quad \text{and} \quad 1 \le i \le n,$$
$$B_j = [t_{j-1}, t_j], \quad h_j = t_j - t_{j-1} \quad \text{and} \quad 1 \le j \le m.$$

Then the lower sum, $L(f, P_1, P_2)$, and the upper sum, $U(f, P_1, P_2)$, are defined by

$$L(f, P_1, P_2) = \sum_{i=1}^{n} \sum_{j=1}^{m} \inf\{f(s, t) : (s, t) \in A_i \times B_j\} l_i h_j,$$

and

$$U(f, P_1, P_2) = \sum_{i=1}^{n} \sum_{j=1}^{m} \sup\{f(s, t) : (s, t) \in A_i \times B_j\} l_i h_j.$$

This notation is rather cumbersome, so we compress part of it. Firstly, we write

$$\inf\{f(s, t) : (s, t) \in A_i \times B_j\} \quad \text{as} \quad \inf f(A_i \times B_j).$$

Then the expression $\inf(A_i \times B_j) l_i h_j$ corresponds to the volume of the *largest* box with base $A_i \times B_j$ which can be placed *under* f. Similarly, $\sup f(A_i \times B_j) l_i h_j$ corresponds to the volume of the *smallest* box with base $A_i \times B_j$ which can be placed *above* f. Also, when there is no danger of confusion, the indices on the sums will be omitted, so that

$$L(f, P_1, P_2) = \sum_{i,j} \inf f(A_i \times B_j) l_i h_j$$

and

$$U(f, P_1, P_2) = \sum_{i,j} \sup f(A_i \times B_j) l_i h_j.$$

Exercises

1. If P_1' and P_2' are obtained from P_1 and P_2 by adding more points, then $L(f, P_1', P_2') \geq L(f, P_1, P_2)$ and $U(f, P_1', P_2') \leq U(f, P_1, P_2)$.

2. If P_1, P_2 and P_1', P_2' are any two partitions then

$$L(f, P_1, P_2) \leq U(f, P_1', P_2').$$

Thus every lower sum is less than or equal to every upper sum.

Definition 11.1 *Let* f *be a bounded, real-valued function on* $D = [a, b] \times [c, d]$. *The lower integral of* f *on* D, *written* $\underline{\int}_D f$, *is the supremum of all lower sums taken over all possible partitions of* $[a, b]$ *and* $[c, d]$. *The upper integral of* f *on* D, *written* $\overline{\int}_D f$, *is the infimum of all upper sums, taken over all possible partitions of* $[a, b]$ *and* $[c, d]$. *If the values of the lower and upper integrals coincide, then* f *is said to be Riemann-integrable The common value is the Riemann integral of* f.

From the exercises above, it follows that

$$\underline{\int}_D f \leq \overline{\int}_D f.$$

If these two values are equal the Riemann integral of f is written as

$$\int_D f \text{ or } \int_D f(s, t) \, d(s, t) \text{ or } \int \int_D f(s, t) \, ds \, dt.$$

Lemma 11.2 *The function* $f : \mathbb{R}^2 \to \mathbb{R}$ *is Riemann-integrable on* $D = [a, b] \times [c, d]$ *if and only if, given* $\epsilon > 0$, *there exist partitions* P_1 *and* P_2 *of* $[a, b]$ *and* $[c, d]$ *respectively such that*

$$U(f, P_1, P_2) - L(f, P_1, P_2) \leq \epsilon.$$

Proof. Suppose the condition in the lemma holds. Then for all $\epsilon > 0$ we have

$$\overline{\int}_D f - \underline{\int}_D f \leq \epsilon,$$

and so we conclude that $\underline{\int}_D f = \overline{\int}_D f$, that is, f is Riemann-integrable.

On the other hand, suppose f is Riemann-integrable on D. Suppose ϵ is positive. Then there exist partitions P_1 of $[a, b]$ and P_2 of $[c, d]$ such that

$$L(f, P_1, P_2) \geq \underline{\int}_D f - \frac{\epsilon}{2} = \int_D f - \frac{\epsilon}{2}.$$

Similarly, we can choose partitions Q_1 of $[a, b]$ and Q_2 of $[c, d]$ such that

$$U(f, Q_1, Q_2) \leq \int_D f + \frac{\epsilon}{2}.$$

Now create new partitions $R_1 = P_1 \cup Q_1$ and $R_2 = P_2 \cup Q_2$. Then by the previous exercises,

$$L(f, R_1, R_2) \geq L(f, P_1, P_2) \geq \int_D f - \frac{\epsilon}{2},$$

and

$$U(f, R_1, R_2) \leq U(f, Q_1, Q_2) \leq \int_D f + \frac{\epsilon}{2}.$$

Hence

$$U(f, R_1, R_2) - L(f, R_1, R_2) \leq \epsilon. \quad \blacksquare$$

If we apply **11.2**, then in the notation of the proof, we can write

$$U(f, R_1, R_2) - L(f, R_1, R_2) = \sum_{i,j} \{\sup f(A_i \times B_j) - \inf f(A_i \times B_j)\} l_i m_j.$$

The expression $\sup f(A_i \times B_j) - \inf f(A_i \times B_j)$ is sometimes called the oscillation of f on $A_i \times B_j$ and is written osc $f(A_i \times B_j)$. The next theorem collects two familiar facts about integration.

Theorem 11.3 *Let f and g be real-valued, Riemann-integrable functions on the region $D = [a, b] \times [c, d]$ in $\mathbb{R}^2$. Then $f + g$ and $f.g$ are Riemann-integrable on D and $\int_D (f + g) = \int_D f + \int_D g$.*

Proof. Since f and g are Riemann-integrable on D, they are bounded there. Suppose M has been chosen so that $|f(s, t)|$ and $|g(s, t)|$ are at most M for all $(s, t) \in D$. To show that $f + g$ and $f.g$ are Riemann-integrable, it will suffice, by **11.2**, to show that for each $\epsilon > 0$ there exist partitions P_1, P_2 of $[a, b]$ and $[c, d]$ respectively such that

$$\sum_{i,j} \text{osc}\,(f + g)(A_i \times B_j) l_i m_j \leq \epsilon,$$

and

$$\sum_{i,j} \text{osc}\,(f.g)(A_i \times B_j)l_i m_j \leq \epsilon.$$

Take $\epsilon > 0$. Since f and g are Riemann-integrable on D, we can find partitions P_1 of $[a, b]$ and P_2 of $[c, d]$ such that

$$\sum_{i,j} \text{osc}\, f(A_i \times B_j)l_i m_j \leq \epsilon \quad \text{and} \quad \sum_{i,j} \text{osc}\, g(A_i \times B_j)l_i m_j \leq \epsilon,$$

by **11.2**. Now for any two points (s, t) and (s', t') in $A_i \times B_j$ we have

$$
\begin{aligned}
(f + g)(s', t') - (f + g)(s, t) &= f(s', t') - f(s, t) + g(s', t') - g(s, t) \\
&\leq \sup f(A_i \times B_j) - \inf f(A_i \times B_j) \\
&\quad + \sup g(A_i \times B_j) - \inf g(A_i \times B_j) \\
&= \text{osc}\, f(A_i \times B_j) + \text{osc}\, g(A_i \times B_j).
\end{aligned}
$$

Hence,

$$
\begin{aligned}
\text{osc}\,(f + g)(A_i \times B_j) &= \sup(f + g)(A_i \times B_j) - \inf(f + g)(A_i \times B_j) \\
&\leq \text{osc}\, f(A_i \times B_j) + \text{osc}\, g(A_i \times B_j),
\end{aligned}
$$

and so

$$
\begin{aligned}
\sum_{i,j} \text{osc}\,(f + g)(A_i \times B_j)l_i m_j &\leq \sum_{i,j} \text{osc}\, f(A_i \times B_j)l_i m_j \\
&\quad + \sum_{i,j} \text{osc}\, g(A_i \times B_j)l_i m_j \\
&\leq 2\epsilon.
\end{aligned}
$$

Thus, by **11.2**, $f + g$ is Riemann-integrable. A slightly more intricate trick is required for the product. Observe that

$$
\begin{aligned}
f(s', t')g(s', t') - f(s, t)g(s, t) &= f(s', t')\{g(s', t') - g(s, t)\} \\
&\quad + g(s, t)\{f(s', t') - f(s, t)\} \\
&\leq |f(s', t')| \text{osc}\, g(A_i \times B_j) \\
&\quad + |g(s, t)| \text{osc}\, f(A_i \times B_j) \\
&\leq M[\text{osc}\, f(A_i \times B_j) + \text{osc}\, g(A_i \times B_j)].
\end{aligned}
$$

This implies that

$$\text{osc}\,(f.g)(A_i \times B_j) \leq M[\text{osc}\, f(A_i \times B_j) + \text{osc}\, g(A_i \times B_j)]$$

and so

$$\sum_{i,j} \operatorname{osc}(f.g)(A_i \times B_j)l_i m_j \leq M\left[\sum_{i,j} \operatorname{osc} f(A_i \times B_j)l_i m_j\right.$$

$$\left. + \sum_{i,j} \operatorname{osc} g(A_i \times B_j)l_i m_j\right]$$

$$\leq 2M\epsilon.$$

Once again **11.2** shows that $f.g$ is Riemann-integrable.

Finally, for any partitions P_1 of $[a, b]$ and P_2 of $[c, d]$ and for any $(s, t) \in A_i \times B_j$, we have

$$f(s, t) + g(s, t) \leq \sup f(A_i \times B_j) + \sup g(A_i \times B_j),$$

and so

$$\sup(f + g)(A_i \times B_j) \leq \sup f(A_i \times B_j) + \sup g(A_i \times B_j).$$

Thus

$$\sum_{i,j} \sup(f + g)(A_i \times B_j)l_i m_j \leq \sum_{i,j} \sup f(A_i \times B_j)l_i m_j$$

$$+ \sum_{i,j} \sup g(A_i \times B_j)l_i m_j,$$

so that

$$U(f + g, P_1, P_2) \leq U(f, P_1, P_2) + U(g, P_1, P_2).$$

Consequently,

$$\overline{\int}_D (f + g) \leq \overline{\int}_D f + \overline{\int}_D g.$$

A similar argument shows that

$$\underline{\int}_D (f + g) \geq \underline{\int}_D f + \underline{\int}_D g,$$

and so

$$\int_D f + \int_D g = \underline{\int}_D f + \underline{\int}_D g$$

$$\leq \underline{\int}_D (f + g) = \int_D (f + g)$$

$$= \overline{\int}_D (f + g)$$

$$\leq \overline{\int}_D f + \overline{\int}_D g$$

$$= \int_D f + \int_D g.$$

These inequalities show that $\int_D (f+g) = \int_D f + \int_D g$. ∎

Exercises

1. Let f be a real-valued function on $D = [a,b] \times [c,d]$. Suppose that given $\epsilon > 0$ there exist real-valued, Riemann-integrable functions on D such that $g \leq f \leq h$ and $\int_D (h-g) \leq \epsilon$. Show that f is Riemann-integrable. Note that $g \leq f \leq h$ means that for all $(s,t) \in D$ we have $g(s,t) \leq f(s,t) \leq h(s,t)$.

One of the central requirements of a decent theory of integration is that there should be a good supply of integrable functions. The next result shows that this is the case for the Riemann integral. The proof relies heavily on results from chapter **6**.

Theorem 11.4 *If* $f : \mathbb{R}^2 \to \mathbb{R}$ *is continuous on* $D = [a,b] \times [c,d]$, *then* f *is Riemann-integrable on* D.

Proof. If f is continuous on D, then f is bounded on D by **6.4**, since D is a compact subset of $\mathbb{R}^2$ by **6.17**. Also, by **6.10**, f is uniformly continuous on D. Now set $M = (b-a)(d-c)$ and take $\epsilon > 0$. Then there is a $\delta > 0$ such that if (s,t) and (s',t') are in D and $\|(s,t) - (s',t')\| < \delta$ then $|f(s,t) - f(s',t')| < \epsilon/M$. Take partitions P_1 of $[a,b]$ and P_2 of $[c,d]$ such that

$$\max_{1 \leq i \leq n} l_i = \max_{1 \leq i \leq n}(s_i - s_{i-1}) \leq \delta \quad \text{and} \quad \max_{1 \leq j \leq m} m_j = \max_{1 \leq j \leq m}(t_j - t_{j-1}) \leq \delta.$$

It then follows that if (s,t) and (s',t') are in $A_i \times B_j = [s_{i-1}, s_i] \times [t_{j-1}, t_j]$ then

$$|f(s,t) - f(s',t')| < \epsilon/M,$$

and so osc $f(A_i \times B_j) \leq \epsilon/M$ for $1 \leq i \leq n$ and $1 \leq j \leq m$. Hence,

$$\begin{aligned}
U(f, P_1, P_2) - L(f, P_1, P_2) &= \sum_{i,j} \text{osc } f(A_i \times B_j) l_i m_j \\
&\leq \sum_{i,j} \frac{\epsilon}{M} l_i m_j \\
&= \frac{\epsilon}{M} \sum_{i,j} l_i m_j \\
&= \frac{\epsilon}{M}(b-a)(d-c) \\
&= \epsilon.
\end{aligned}$$

It now follows from **11.2** that f is Riemann-integrable. ∎

It is perhaps worth recalling at this point that an essential feature in the definition of the Riemann integral is the boundedness of the function. This is also true when defining the Riemann integral of a function from $\mathbb{R}$ to $\mathbb{R}$. Consequently, the function $f : \mathbb{R} \to \mathbb{R}$ given by

$$f(s) = \begin{cases} 0 & s = 0 \\ |s|^{-1/2} & s \neq 0 \end{cases}$$

is not Riemann-integrable on $[0, 1]$. Of course, we can extend our notion of integration by declaring

$$\int_{[0,1]} f(s)\, ds = \lim_{\alpha \to 0^+} \int_{[\alpha,1]} f(s)\, ds.$$

Such integrals are often called improper Riemann integrals.

It is also important to be aware of the fact that other definitions of integration are possible. These define volumes just as well as the Riemann integral, but are often superior in other respects.

The ingredient that is missing in a workable theory of integration is a useful technique for evaluating Riemann integrals. Familiarity with the so-called fundamental theorem of calculus in one variable will be assumed. Thus, if $[a, b]$ is a closed interval in $\mathbb{R}$ and $f : [a, b] \to \mathbb{R}$ possesses an *antiderivative* F, i.e., the function $F : [a, b] \to \mathbb{R}$ has the property that $F'(s) = f(s)$ for all $s \in [a, b]$, then

$$\int_{[a,b]} f(s)\, ds = F(b) - F(a).$$

The following result provides the key to the mechanics of integration.

Theorem 11.5 *Let $D = [a, b] \times [c, d]$ and suppose f is a real-valued, Riemann-integrable function on D. For $s \in [a, b]$ define $\underline{F}$ and $\overline{F}$ from $[a, b]$ to $\mathbb{R}$ by*

$$\underline{F}(s) = \underline{\int}_{[c,d]} f(s, t)\, dt \quad and \quad \overline{F}(s) = \overline{\int}_{[c,d]} f(s, t)\, dt.$$

Then $\underline{F}$ and $\overline{F}$ are Riemann-integrable on $[a, b]$ and

$$\int_{[a,b]} \underline{F}(s)\, ds = \int_{[a,b]} \overline{F}(s)\, ds = \int\int_D f(s, t)\, ds\, dt.$$

Proof. Take partitions P_1 of $[a, b]$ and P_2 of $[c, d]$. Then

$$L(f, P_1, P_2) = \sum_{i,j} \inf f(A_i \times B_j) l_i m_j.$$

Now for a fixed $s \in [a, b]$, write $f_s(t) = f(s, t)$. Hence, for each s in $[a, b]$, f_s is a bounded, real-valued function on $[c, d]$. If $s \in A_i$ then

$$\inf f(A_i \times B_j) \leq \inf f_s(B_j)$$

since $\{s\} \times B_j \subset A_i \times B_j$. Hence, for $s \in A_i$,

$$
\begin{aligned}
\sum_j \inf f(A_i \times B_j) m_j & \leq \sum_j \inf f_s(B_j) m_j \\
& = L(f_s, P_2) \\
& \leq \underline{\int}_{[c,d]} f_s \\
& = \underline{F}(s).
\end{aligned}
$$

Now this inequality holds for each $s \in A_i$, and so

$$\sum_j \inf f(A_i \times B_j) m_j \leq \inf\{\underline{F}(s) : s \in A_i\} = \inf \underline{F}(A_i).$$

Multiplying this inequality by l_i and then summing over i gives

$$\sum_{i,j} \inf f(A_i \times B_j) l_i m_j \leq \sum_i \inf \underline{F}(A_i) l_i = L(\underline{F}, P_1),$$

so that $L(f, P_1, P_2) \leq L(\underline{F}, P_1)$. Now using the definition of the lower integral (**11.1**),

$$L(f, P_1, P_2) \leq L(\underline{F}, P_1) \leq \underline{\int}_{[a,b]} \underline{F}.$$

In the above inequality, take a supremum over all partitions P_1 and P_2. Since f is Riemann-integrable, this gives

$$\int_D f \leq \underline{\int}_{[a,b]} \underline{F}.$$

A similar argument shows that $\overline{\int}_{[a,b]} \overline{F} \leq \int_D f$ and so

$$\int_D f \leq \underline{\int}_{[a,b]} \underline{F} \leq \overline{\int}_{[a,b]} \underline{F} \leq \overline{\int}_{[a,b]} \overline{F} \leq \int_D f.$$

This shows that $\underline{F}$ is Riemann-integrable on $[a, b]$ and that

$$\int_{[a,b]} \underline{F} = \int_D f.$$

Also,

$$\int_D f \leq \int_{[a,b]} \underline{F} \leq \overline{\int}_{[a,b]} \overline{F} \leq \overline{\int}_{[a,b]} \overline{F} \leq \int_D f$$

shows the same for $\overline{F}$. ∎

The following is a direct consequence of **11.5** and is easier to apply.

Corollary 11.6 *Let f be a real-valued, Riemann-integrable function on $D = [a, b] \times [c, d]$. If $\int_c^d f(s, t)\, dt$ exists for all $s \in [a, b]$ and $\int_a^b f(s, t)\, ds$ exists for all $t \in [c, d]$, then*

$$\int_a^b \left(\int_c^d f(s, t)\, dt \right) ds = \int_c^d \left(\int_a^b f(s, t)\, ds \right) dt = \int_D f.$$

Example Suppose $f : [0, 1] \times [-1, 2] \to \mathbb{R}$ is given by $f(s, t) = st^2$. Then

$$\int_{-1}^2 f(s, t)\, dt = \int_{-1}^2 st^2\, dt = 3s, \quad s \in [0, 1]$$

while

$$\int_0^1 f(s, t)\, ds = \int_0^1 st^2\, ds = \frac{t^2}{2}, \quad t \in [-1, 2].$$

Hence

$$\int_D f = \int_0^1 3s\, ds = \frac{3}{2} = \int_{-1}^2 \frac{t^2}{2}\, dt.$$

Exercises

1. Evaluate $\int_D st\, ds\, dt$ where $D = [-1, 1] \times [0, 2]$.

2. Evaluate

$$\int_0^1 \left(\int_0^1 \frac{s^2 - t^2}{(s^2 + t^2)^2}\, ds \right) dt \quad \text{and} \quad \int_0^1 \left(\int_0^1 \frac{s^2 - t^2}{(s^2 + t^2)^2}\, dt \right) ds.$$

Hint: it might help to observe, in the notation of chapter **9**, that if

$$g(s, t) = \frac{s}{s^2 + t^2} \quad \text{then} \quad (D_1 g)(s, t) = \frac{t^2 - s^2}{(s^2 + t^2)^2}.$$

Suppose $f : \mathbb{R}^2 \to \mathbb{R}$ is defined by

$$f(s,t) = \begin{cases} (s^2 - t^2)/(s^2 + t^2) & s \neq 0 \\ 0 & s = 0. \end{cases}$$

Is f Riemann-integrable on $[0,1] \times [0,1]$?

The next two results concern the special case when the function being integrated involves partial derivatives in a simple way. In particular, **11.8** establishes a result which was mentioned in the discussions of chapter **9**.

Lemma 11.7 *Suppose f is a real-valued function on $\mathbb{R}^2$ such that $D_{12}f$ and $D_{21}f$ are continuous on $D = [a,b] \times [c,d]$. Then*

$$\int_D D_{12}f = \int_D D_{21}f = f(b,d) - f(a,d) - f(b,c) + f(a,c).$$

Proof. By **11.4**, $D_{12}f$ and $D_{21}f$ are Riemann-integrable. Now, for example, if we set $D_1 f = g$ and define $g_s(t) = g(s,t)$ for each $(s,t) \in D$, then

$$(D_{21})f(s,t) = \frac{d}{dt}[g_s(t)].$$

Hence,

$$\int_c^d (D_{21}f)(s,t)\,dt = g_s(d) - g_s(c) = (D_1 f)(s,d) - (D_1 f)(s,c),$$

for each $s \in [a,b]$. Thus

$$\begin{aligned} \int_D D_{21}f &= \int_a^b \left(\int_c^d (D_{21}f)(s,t)\,dt \right) ds \\ &= \int_a^b [(D_1 f)(s,d) - (D_1 f)(s,c)]\,ds \\ &= f(b,d) - f(a,d) - f(b,c) + f(a,c). \end{aligned}$$

The argument is similar for $\int_D D_{12}f$. ∎

Theorem 11.8 *Suppose $f : \mathbb{R}^2 \to \mathbb{R}$ has $D_{12}f$ and $D_{21}f$ continuous in some open ball $B_\delta((s_0, t_0))$, where $\delta > 0$. Then $(D_{12}f)(s_0, t_0) = (D_{21}f)(s_0, t_0)$.*

Proof. Suppose that $(D_{12}f)(s_0, t_0) \neq (D_{21})f(s_0, t_0)$. Then we can write

$$(D_{12}f)(s_0, t_0) = (D_{21})f(s_0, t_0) + 2\epsilon,$$

where $\epsilon \neq 0$. Assume $\epsilon > 0$. Since both $D_{12}f$ and $D_{21}f$ are continuous, there must exist $a < s_0 < b$ and $c < t_0 < d$ such that

$$(D_{12}f)(s, t) > (D_{21}f)(s, t) + \epsilon,$$

for all $(s, t) \in D = [a, b] \times [c, d]$. Then

$$\int_D D_{12}f > \int_D D_{21}f + \epsilon(b - a)(d - c).$$

But from **11.7** we see that $\int_D D_{12}f = \int_D D_{21}f$ and so this contradiction establishes the result. ■

One of the attractive features of the Riemann integral as we have defined it is that integration may be considered over sets other than rectangles in $\mathbb{R}^2$. This generalisation is quite easy to achieve with the aid of characteristic functions.

Definition 11.9 *Let E be a set in $\mathbb{R}^2$. Then the characteristic function of E, denoted by χ_E is the function from $\mathbb{R}^2$ into $\mathbb{R}$ defined by*

$$\chi_E(s, t) = \begin{cases} 0 & (s, t) \notin E \\ 1 & (s, t) \in E \end{cases}.$$

Note that characteristic functions can be defined in a much more general setting than subsets of $\mathbb{R}^2$. Not all characteristic functions are Riemann-integrable, as can be seen from the exercises at the end of the chapter.

Definition 11.10 *Let E be a subset of $D = [a, b] \times [c, d]$, such that χ_E is Riemann-integrable. If f is Riemann-integrable on D, then we define*

$$\int_E f = \int_D (f\chi_E).$$

Observe in **11.10** that $f\chi_E$ is a function which agrees with f on E and has value 0 on $D \setminus E$. Also **11.3** says that since f and χ_E are both Riemann-integrable on D, the function $f\chi_E$ is also Riemann-integrable on D. This remark establishes that **11.10** is a workable definition. For it to be a useful definition, there has to be some reasonable criterion for judging whether χ_E is Riemann-integrable for a given E in $\mathbb{R}^2$.

Lemma 11.11 *Let* $h : [a, b] \to \mathbb{R}$ *be a Riemann-integrable function with* $c \le h(s) \le d$ *for all* s *in* $[a, b]$. *Set*

$$E = \{(s, t) : a \le s \le b, \ c \le t \le h(s)\}.$$

Then χ_E *is Riemann-integrable, and* $\int \chi_E = \int_a^b h$.

Proof. Take $\epsilon > 0$. Then since h is Riemann-integrable, there is a partition

$$P_1 \ : \ a < s_0 < s_1 < \ldots < s_n = b$$

such that

$$\sum_i \sup h(A_i) l_i \le \int h + \epsilon.$$

Set $\epsilon' = \epsilon/(b - a)$ and $M_i = \sup h(A_i)$. At this point it may be helpful to refer to figure 11.5. Now let P_2 be the partition of $[c, d]$ obtained by using the numbers $M_i + \epsilon'$ as division points. Then an upper bound for the total area of rectangles that intersect E is

$$\sum_i (M_i + \epsilon')(s_i - s_{i-1}).$$

Also,

$$\begin{aligned}
\sum_i (M_i + \epsilon')(s_i - s_{i-1}) &= \sum_i M_i l_i + \sum_i \epsilon' l_i \\
&\le \int h + \epsilon + \epsilon'(b - a) \\
&= \int h + 2\epsilon.
\end{aligned}$$

Now observe that

$$\sum_i (M_i + \epsilon')(s_i - s_{i-1}) = U(\chi_E, P_1, P_2),$$

so that

$$\overline{\int} \chi_E \le U(\chi_E, P_1, P_2) \le \int h + 2\epsilon.$$

Now this inequality is true for all $\epsilon > 0$ and so $\overline{\int} \chi_E \le \int h$. A similar argument shows that $\underline{\int} \chi_E \ge \int h$ and so

$$\int h \le \underline{\int} \chi_E \le \overline{\int} \chi_E \le h.$$

Thus $\underline{\int} \chi_E = \overline{\int} \chi_E = \int h$, and so χ_E is Riemann-integrable with $\int \chi_E = \int h$. ∎

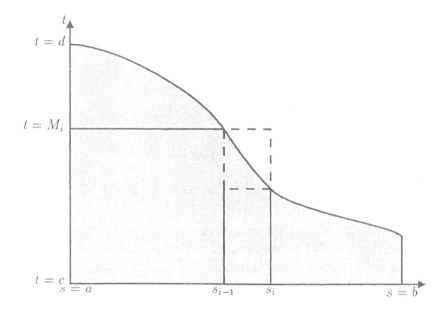

Figure 11.5: Illustration of proof of 10.11.

There is a simple (but unfortunately non-rigorous) argument which shows that **11.11** is true. From the definition of the univariate Riemann integral, we know that

$$\text{Area of } E = \int_a^b h.$$

Now χ_E is a function of height 1 above E (if we consider the usual setting where χ_E is graphed in $\mathbf{R}^3$). Hence

$$\int_a^b h = \text{Area of } E = \text{Volume under } \chi_E = \int_D \chi_E.$$

The misssing ingredient from this argument is whether the function χ_E is Riemann-integrable.

Now the results concerning Riemann integrals on rectangles can be translated, without the need for proof, to sets $E \subset \mathbf{R}^2$ such that χ_E is Riemann-integrable. The most useful is **11.5**.

Theorem 11.12 *Let* $f : \mathbf{R}^2 \to \mathbf{R}$ *be Riemann-integrable on the domain* $D = [a, b] \times [c, d]$. *Let* $h : [a, b] \to \mathbf{R}$ *be Riemann-integrable with*

$c \leq h(s) \leq d$ *for* $s \in [a, b]$. *If* $\int_c^d f(s, t)\, dt$ *exists for all* $s \in [a, b]$, *then* f *is Riemann-integrable on* $E = \{(s, t) : a \leq s \leq b,\ c \leq t \leq h(s)\}$ *and*

$$\int_E f = \int_a^b \left(\int_c^{h(s)} f(s, t)\, dt \right) ds.$$

We conclude this chapter with a simple application of Riemann integration. Suppose $D = [a, b] \times [a, b]$ and $f : [a, b] \to \mathbb{R}$, $g : [a, b] \to \mathbb{R}$ are Riemann-integrable. Suppose $F : [a, b] \to \mathbb{R}$ and $G : [a, b] \to \mathbb{R}$ can be defined by

$$F(s) = \int_a^s f \quad \text{and} \quad G(t) = \int_a^t g.$$

Define $z : D \to \mathbb{R}$ by $z(s, t) = f(s)g(t)$, and $E \subset \mathbb{R}^2$ by

$$E = \{(s, t) : a \leq s \leq b,\ a \leq t \leq s\}.$$

Then **11.11** shows that χ_E is Riemann-integrable. Also

$$\int_a^b z(s, t)\, dt = \int_a^b f(s)g(t)\, dt = f(s) \int_a^b g(t)\, dt,$$

and so $\int_a^b z(s, t)\, dt$ exists for each $s \in [a, b]$. Lastly, z is Riemann-integrable on D, and so

$$\int_E z = \int_a^b \left(\int_a^s f(s)g(t)\, dt \right) ds = \int_a^b f(s) \left(\int_a^s g(t)\, dt \right) ds.$$

Thus, by the fundamental theorem of calculus,

$$\begin{aligned}
\int_E z &= \int_a^b f(s) \left(\int_a^s g(t)\, dt \right) ds \\
&= \int_a^b f(s)\{G(s) - G(a)\}\, ds \\
&= \int_a^b f(s)G(s)\, ds - G(a) \int_a^b f(s)\, ds \\
&= \int_a^b f(s)G(s)\, ds - G(a)[F(b) - F(a)].
\end{aligned}$$

Now the same process may be carried out with the reverse order of integration, so that

$$\int_E z = \int_a^b g(t) \left(\int_t^b f(s)\, ds \right) dt$$

$$= \int_a^b g(t)\{F(b) - F(t)\}\,dt$$

$$= -\int_a^b g(t)F(t)\,dt + F(b)\int_a^b g(t)\,dt$$

$$= -\int_a^b g(t)F(t)\,dt + F(b)[G(b) - G(a)].$$

Now equating the two formulae for $\int_E z$ gives

$$\int_a^b f(s)G(s)\,ds = F(b)G(b) - F(a)G(a) - \int_a^b g(s)F(s)\,ds.$$

This is exactly the formula for univariate integration by parts!

Exercises

1. Let $D = [-1, 1] \times [0, 2]$ and define $f : D \to \mathbb{R}^2$ by $f(s, t) = st$. Find the integral of f on D and on the triangles in D below each of the two diagonals.

2. Let f, g be defined on $\mathbb{R}^+$ in such a way that f and g are Riemann-integrable on each interval $[0, s]$, $s > 0$. Set $F(s) = \int_0^s f(s)\,ds$ and $G(s) = \int_0^s g(s)\,ds$, $s \in \mathbb{R}^+$. Define the convolution $f * g$ by

$$(f * g)(s) = \int_0^s f(s - t)g(t)\,dt, \quad s \geq 0.$$

By assuming that the theorem on repeated intergrals is applicable, prove that

$$\int_0^a (f * g) = (F * g)(a), \quad a \geq 0,$$

and also that $(f * g) * h = f * (g * h)$.

3. Show that in the previous question the interchange of repeated integrals is indeed possible. That is, if $f : [-a, a] \to \mathbb{R}$ is Riemann-integrable, then $\hat{f}$ is Riemann-integrable on $D = [0, a] \times [0, a]$ where $\hat{f}(s, t) = f(s - t)$.

4. By considering the integral of $f : \mathbb{R}^2 \to \mathbb{R}$ defined by $f(s, t) = e^{st}$ on a suitable triangle, prove that

$$2\int_1^a \frac{1}{s} e^{s^2}\,ds = \int_1^a \frac{1}{s}(e^{as} + e^s)\,ds.$$

5. By considering the integral of $f : \mathbb{R}^2 \to \mathbb{R}$ defined by $f(s,t) = se^{st}$ on a suitable triangle, prove that

$$2 \int_0^a e^{s^2} ds = \int_0^a \frac{1}{s^2}(e^{s^2} - 1)\, ds + \frac{1}{a}(e^{a^2} - 1).$$

6. Set $A = (a,b) \times (c,d)$. Suppose $f : A \to \mathbb{R}$ is continuous and $(0,0) \in A$. Define $F : A \to \mathbb{R}$ by

$$F(s,t) = \int_0^s \int_0^t f(u,v)\, du\, dv.$$

Show that $(D_{12}F)(s,t) = (D_{21}F)(s,t) = f(s,t)$ for all $(s,t) \in A$.

7. The function $f : [0,1] \to \mathbb{R}$ defined by

$$f(s) = \begin{cases} 1 & s \text{ rational} \\ 0 & s \text{ irrational,} \end{cases}$$

is called Dirichlet's discontinuous function. Show that f is not Riemann-integrable.

8. Use the previous exercise to contruct a set E in $\mathbb{R}^2$ such that its characteristic function χ_E is not Riemann-integrable.

12

Two hard results

Most of the results covered in this book are relatively straightforward. That is, they rely on a small number of concepts, and their proofs are short and self-contained. We conclude with two very important results, both of which have rather difficult proofs. It is interesting that both these results often feature in elementary calculus courses, so that the student is using tools which are far more sophisticated in their content and proof than can be understood from such a viewpoint.

The first theorem concerns the change of variable process in integration. We shall not attempt to provide the most general theorem (see for example Rosenlicht [4]), since that would obscure the main ideas of the proof in a mass of details. Instead, following the tradition of chapter 11, we shall consider integration of functions $f : \mathbb{R}^2 \to \mathbb{R}$ over some domain in $\mathbb{R}^2$. We will also limit ourselves to only a small selection of changes of variable. Recall that if ϕ is a real-valued function on the interval $[a, b]$ such that ϕ' is continuous on $[a, b]$ and f is a real-valued, continuous function on $\phi([a, b])$, then the univariate change of variable theorem states that

$$\int_{\phi(a)}^{\phi(b)} f = \int_a^b (f \circ \phi)\phi' = \int_a^b f(\phi(s))\phi'(s)\, ds.$$

Thus, for example, the integral

$$\int_0^2 2s(s^2 + 1)^3\, ds$$

can be recognised as

$$\int_0^2 \phi'(s)[(\phi(s)]^3\, ds,$$

where ϕ is defined by $\phi(s) = s^2 + 1$. Now via the theorem on change of variable,

$$\int_0^2 2s(s^2 + 1)^3 \, ds = \int_{\phi(0)}^{\phi(2)} x^3 \, dx = \int_1^5 x^3 dx = 156.$$

It is the analogue of this process for integration of bivariate functions to which we now turn. The strategy is to consider initially transformations ϕ of the region D in $\mathbb{R}^2$ which are of the form $(s, t) \mapsto (s, h(s, t))$, where h is a real-valued function on $\mathbb{R}^2$. Then more general transformations ϕ can be obtained by compositions of such mappings.

Lemma 12.1 *Let*

$$A = \{(s, t) : a \leq s \leq b, \ p_1(s) \leq t \leq p_2(s)\},$$

where p_1 and p_2 are continuous, real-valued mappings on $[a, b]$. Let h be a real-valued, continuous function on A such that $D_2 h$ exists and is continuous on A. Let $\sigma : A \to \mathbb{R}^2$ be defined by $\sigma(s, t) = (s, h(s, t))$. If $(D_2 h)(s, t) \geq 0$ for all $(s, t) \in A$ and f is continuous on $\sigma(A)$, then

$$\int_{\sigma(A)} f = \int_A (f \circ \sigma)(D_2 h).$$

Proof. For a fixed s in $[a, b]$ the mapping $t \mapsto h(s, t)$ is continuous on $[p_1(s), p_2(s)]$. Since $(D_2 h)(s, t) \geq 0$ for all $t \in [p_1(s), p_2(s)]$, this mapping is also increasing and so maps the interval $[p_1(s), p_2(s)]$ onto the interval $[q_1(s), q_2(s)]$ where $q_1(s) = h(s, p_1(s))$ and $q_2(s) = h(s, p_2(s))$. Thus the region $\sigma(A)$ can be written as

$$\sigma(A) = \{(s, t) : a \leq s \leq b, \ q_1(s) \leq t \leq q_2(s)\}.$$

Also, since f, σ and $D_2 h$ are continuous functions, the mapping $(f \circ \sigma)(D_2 h)$ is a continuous function from A into $\mathbb{R}^2$. Hence, by **11.4**, this function is Riemann-integrable on A. Define

$$I(s) = \int_{p_1(s)}^{p_2(s)} f(s, h(s, t))(D_2 h)(s, t) \, dt, \quad \text{for all } s \in [a, b].$$

Now fix s in $[a, b]$ and write

$$\left. \begin{array}{rcl} \tilde{f}(t) & = & f(s, t) \\ \tilde{h}(t) & = & h(s, t) \end{array} \right\} \quad \text{for } p_1(s) < t < p_2(s).$$

Then $I(s)$ can be written

$$I(s) = \int_{p_1(s)}^{p_2(s)} \tilde{f}(\tilde{h}(t))\tilde{h}'(t)\,dt.$$

The usual univariate change of variable theorem may be applied to deduce that

$$I(s) = \int_{\tilde{h}(p_1(s))}^{\tilde{h}(p_2(s))} \tilde{f}(t)\,dt = \int_{q_1(s)}^{q_2(s)} f(s,t)\,dt.$$

Now by the result on repeated integrals (see **11.12**), we obtain finally,

$$
\begin{aligned}
\int_A (f \circ \sigma)(D_2 h) &= \int_a^b \left(\int_{p_1(s)}^{p_2(s)} f(s, h(s,t))(D_2 h)(s,t)\,dt \right) ds \\
&= \int_a^b \left(\int_{q_1(s)}^{q_2(s)} f(s,t)\,dt \right) ds \\
&= \int_{\sigma(A)} f. \qquad \blacksquare
\end{aligned}
$$

There is, of course, a similar statement for transformations of the form

$$\tau(s,t) = (g(s,t), t).$$

We will give this in a moment in notation which is a little more suited to the coming discussion. Recall from chapter **10** that if the mapping $\phi : \mathbb{R}^2 \to \mathbb{R}^2$ is written as $\phi(s,t) = (\phi_1(s,t), \phi_2(s,t))$, then $D\phi$, if it exists, is the linear mapping from $\mathbb{R}^2$ to itself given by

$$D\phi(s,t) = \begin{pmatrix} (D_1\phi_1)(s,t) & (D_1\phi_2)(s,t) \\ (D_2\phi_1)(s,t) & (D_2\phi_2)(s,t) \end{pmatrix}.$$

We shall write $J_\phi(s,t)$ for $\det(D\phi(s,t))$. The mapping σ in **12.1** has

$$D\sigma(s,t) = \begin{pmatrix} 1 & (D_1 h)(s,t) \\ 0 & (D_2 h)(s,t) \end{pmatrix},$$

so that $J_\sigma(s,t) = (D_2 h)(s,t)$. Thus we could have expressed the conclusion of **12.1** as

$$\int_{\sigma(A)} f = \int_A (f \circ \sigma) J_\sigma.$$

There now follows, without proof, a reformulation of **12.1** for transformations of the form $\tau(s,t) = (g(s,t), t)$.

Lemma 12.2 *Let*

$$B = \{(s,t) : u \le t \le v, \ q_1(t) \le s \le q_2(t)\},$$

where q_1 and q_2 are continuous, real-valued mappings on $[u,v]$. Let g be a continuous, real-valued function on B, and suppose D_1g exists, is continuous and is non-negative on B. Let $\tau(s,t) = (g(s,t),t)$ so that $J_\tau = D_1g$. If f is a continuous, real-valued function on B, then

$$\int_{\tau(B)} f = \int_B (f \circ \tau) J_\tau.$$

We now come to the main result about change of variables.

Theorem 12.3 *Let $E = [a,b] \times [c,d]$ and $E' = [a',b'] \times [c',d']$, where $a' < a < b < b'$ and $c' < c < d < d'$. Let $\phi = (\phi_1, \phi_2)$ be a mapping from $\mathbb{R}^2$ into itself, having continuous partial derivatives throughout E', and such that $J_\phi(s,t) > 0$ and $(D_2\phi_2)(s,t) > 0$ for all $(s,t) \in E'$. Suppose that the univariate functions*

$$p_1(s) = \phi_2(s,c) \quad and \quad p_2(s) = \phi_2(s,d)$$

are strictly monotonic on $[a,b]$. If f is a continuous, real-valued function on E, then

$$\int_{\phi(E)} f = \int_E (f \circ \phi) J_\phi.$$

Proof. As indicated prior to the statement of the theorem, the idea is to view ϕ as the composition of two mappings, each of which shifts only one of the coordinates. The first mapping, σ, is easily defined as

$$\sigma(s,t) = (s, \phi_2(s,t)), \quad (s,t) \in E'.$$

A glance at **12.1** will show that the mapping σ which we have just defined enjoys the same properties on E' as the mapping designated by the same symbol in **12.1** does on the region designated there by A. Thus we can conclude that

$$\int_{\sigma(E)} h = \int_E (h \circ \sigma) J_\sigma \tag{†}$$

for any function h which is continuous on $\sigma(E)$. Our second mapping, τ, is defined by

$$\tau(s, \phi_2(s,t)) = (\phi_1(s,t), \phi_2(s,t)), \quad (s,t) \in E'.$$

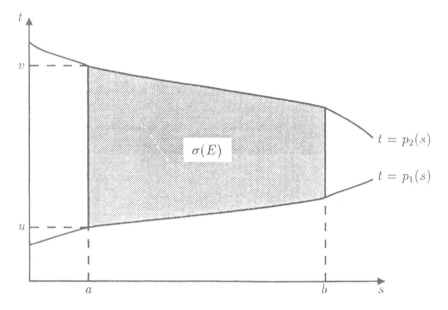

Figure 12.1: A possible configuration of $\sigma(E)$.

It is clear that $\phi = \tau \circ \sigma$. We want to be able to apply **12.2** to τ, and this is where the delicate arguments occur. We need to view τ as a mapping from $\sigma(E)$ onto $(\tau \circ \sigma)(E)$, that is, onto $\phi(E)$. We can write

$$\sigma(E) = \{(s,t) : a \leq s \leq b,\ p_1(s) \leq t \leq p_2(s)\}.$$

By hypothesis, p_1 and p_2 are strictly monotone, and so $\sigma(E)$ can also be written as

$$\sigma(E) = \{(s,t) : u \leq t \leq v,\ q_1(t) \leq s \leq q_2(t)\},$$

where $q_1 = p_1^{-1}$, $q_2 = p_2^{-1}$ and u and v are the appropriate real numbers. For example, $u = \min\{p_1(a), p_2(a)\}$. Figure 12.1 illustrates one possible situation. (There are four possibilities.) This alternative representation of $\sigma(E)$ is ideally suited to an application of **12.2**, and all that remains is to check that τ has the desired properties. Since $(D_2\phi_2)(s,t) > 0$ for all (s,t) in E', the mapping $t \mapsto \phi_2(s,t)$ is strictly increasing on $[c', d']$, for each s in $[a', b']$, and so σ is injective on E'. Hence, σ^{-1} is a continuous function. Thus τ can be written as $\phi \circ \sigma^{-1}$ on $\sigma(E')$. By **2.9**, τ is a continuous mapping on $\sigma(E')$ having the form

$$\tau(s,t) = (\phi_1(s,z), t),$$

where $z = \phi_2(s,t)$. We now claim that σ^{-1} is differentiable at all interior points of $\sigma(E')$. To see this, take any interior point $y \in \sigma(E')$. Let B be an open set in $\sigma(E')$ which contains y. Let A be the set $\sigma^{-1}(B)$. Since σ is continuous, A is open by **3.7**, and σ is then a bijection of A onto B, whose inverse function σ^{-1} is continuous. Since $J_\phi(s,t) = (D_2\phi_2)(s,t) > 0$ for all (s,t) in E', it follows that the mapping $D\sigma(z)$ has a continuous inverse for each z in $\sigma^{-1}(B)$. It follows from **10.6** that σ^{-1} is differentiable at y as required. Now, using **10.3** and the remarks following that result, the chain rule applied to $\phi = \tau \circ \sigma$ gives

$$D\phi = D(\tau \circ \sigma) = (D\tau \circ \sigma)D\sigma,$$

and this equality is certainly valid over the whole of E. From this we conclude that

$$J_\phi = \det D\phi = (\det(D\tau) \circ \phi)(\det D\phi) = (J_\tau \circ \sigma)J_\sigma. \tag{$\ddagger$}$$

Both J_ϕ and J_σ are continuous on E, and $J_\sigma = D_2\phi_2$, so that J_ϕ is positive on E. Hence $J_\tau \circ \sigma$ must be continuous on E, so that J_τ is continuous on $\sigma(E)$. Furthermore, since $J_\phi(s,t) > 0$ for $(s,t) \in E'$, we have that J_τ is a positive function on $\sigma(E)$. This establishes that τ possesses all the properties required of it on $\sigma(E)$ for **12.2** to be applicable. By another application of ($\ddagger$),

$$\begin{aligned}
(f \circ \phi)J_\phi &= (f \circ \tau \circ \sigma)(J_\tau \circ \sigma)J_\sigma \\
&= \{[(f \circ \tau)J_\tau] \circ \sigma\}J_\sigma \\
&= (h \circ \sigma)J_\sigma,
\end{aligned}$$

where $h = (f \circ \tau)J_\tau$. Observe that h is continuous on $\sigma(E)$. Now applying **12.1**, we obtain

$$\int_E (h \circ \sigma)J_\sigma = \int_{\sigma(E)} h = \int_{\sigma(E)} (f \circ \tau)J_\tau.$$

By applying **12.2**, we obtain

$$\int_{\sigma(E)} (f \circ \tau)J_\tau = \int_{(\tau\circ\sigma)(E)} f = \int_{\phi(E)} f,$$

which leads to the desired conclusion

$$\int_{\phi(E)} f = \int_E (h \circ \sigma)J_\sigma = \int_E (f \circ \phi)J_\phi. \qquad \blacksquare$$

The second result concerns implicit functions, and will be treated in a very general way. Suppose X and Y are normed linear spaces and f is a mapping from $X \times Y$ onto Y, such that the point (a, b) in $X \times Y$ is mapped to the zero vector in Y. The equation

$$f(x, y) = \theta_Y$$

is said to define each variable implicitly in terms of the other. The problem of implicit functions is to solve the equation explicitly for one of the variables in terms of the other. That is, we seek a function g defined on a subset of X and having values in Y, such that $g(a) = b$ and

$$f(x, g(x)) = \theta_Y$$

for all values of x in the domain of g. Roughly speaking, this amounts to 'solving' for the variable y as $y = g(x)$.

Example Consider the function $f : \mathbb{R}^2 \to \mathbb{R}$ defined by

$$f(s, t) = 5s^2 - t - 4.$$

Then $f(1, 1) = 0$. By rearranging $5s^2 - t - 4 = 0$ as $t = 5s^2 - 4$ and defining $g : \mathbb{R} \to \mathbb{R}$ by $g(s) = 5s^2 - 4$ we see that $g(1) = 1$. Also

$$
\begin{aligned}
f(s, g(s)) &= 5s^2 - g(s) - 4 \\
&= 5s^2 - (5s^2 - 4) - 4 \\
&= 0
\end{aligned}
$$

for all $s \in \mathbb{R}$.

This example has two features. Firstly, the choice of g is unique, and we shall want this property to hold in our genreal treatment. Secondly, g is defined throughout $\mathbb{R}$. In general, we shall not be so demanding – it will be quite satisfactory if we can only define g over a subset of the appropriate space.

Without further conditions on f, the problem may well be insoluble. Consider $f : \mathbb{R}^2 \to \mathbb{R}$ defined by $f(s, t) = s^2 - t^2$. At the point $(a, b) = (0, 0)$ there are many choices for the function g. An obvious one is $g(s) = s$. A less obvious choice is

$$
g(s) = \begin{cases} s & s \text{ rational} \\ -s & s \text{ irrational.} \end{cases}
$$

In this example, even the demand that g be continuous admits two possibilities, $g(s) = s$ and $g(s) = -s$. Such a situation is not usually desirable – we often want to know that the explicit solution for y in terms of x is unique. Further examples are given in the exercises which show that the conditions needed to ensure a unique, continuous function g must be quite subtle. Loosely speaking, the condition needed on f are its differentiablity plus some assumption on its behaviour with respect to the second variable y.

The statement and proof of the theorem make heavy use of the notation established in chapter **10**. In addition, the symbol θ will be greatly overworked, since it is used to represent the origin in both X and Y. The intended meaning can always be deduced from the context.

If X and Y are normed linear spaces, then $X \times Y$ can be made into a normed linear space in many ways. The variety comes via the definition of the norm. We shall stick to one definition, viz:

$$\|(x,y)\| = \max\{\|x\|, \|y\|\}, \quad (x \in X, y \in Y).$$

Theorem 12.4 *Let X and Y be normed linear spaces, E an open set in $X \times Y$ and let $(a, b) \in E$. Suppose that $f : E \to Y$ belongs to $C^1(E)$, that $f(a, b) = \theta$ and that the linear mapping defined by*

$$My = Df(a,b)(\theta, y), \qquad \text{for all } y \in Y,$$

is a bijection of Y onto Y. Then the following hold.
(i) There is an open set W in X such that $a \in W$, and a unique function $g : W \to Y$ such that $g \in C^1(W)$, $b = g(a)$ and

$$f(x, g(x)) = \theta \quad \text{for all } x \text{ in } W.$$

(ii) There is an open set U in $X \times Y$ such that $(a, b) \in U$ and such that the pair (x, y) in U satisfies $f(x, y) = \theta$ if and only if $y = g(x)$ for x in W.

Proof. (i) We will assume $a = b = \theta$, since this involves no loss of generality. Define $h : E \to X \times Y$ by

$$h(x, y) = (x, f(x, y)), \qquad (x, y) \in E.$$

It follows from exercise **3** of chapter **10** that $h \in C^1(E)$, and that

$$Dh(x, y)(u, v) = (u, Df(x, y)(u, v))$$

for $(x, y) \in E$ and $(u, v) \in X \times Y$. Now we claim that $Dh(\theta, \theta)$ is an invertible mapping. To see this, let $L \in \mathcal{B}(X, Y)$ be the mapping defined by

$$Lx = Df(\theta, \theta)(x, \theta), \quad \text{for all } x \in X.$$

Using the linearity of $Df(\theta, \theta)$, we can write

$$
\begin{aligned}
Df(\theta, \theta)(x, y) &= Df(\theta, \theta)\{(x, \theta) + (\theta, y)\} \\
&= Df(\theta, \theta)(x, \theta) + Df(\theta, \theta)(\theta, y) \\
&= Lx + My.
\end{aligned}
$$

Since M is a bijection of Y onto Y by hypothesis, M is invertible. Consider the linear mapping K on $X \times Y$ defined by

$$K(x, z) = (x, M^{-1}(z - Lx)).$$

Then, for any $(x, y) \in X \times Y$,

$$
\begin{aligned}
[K \circ Dh(\theta, \theta)](x, y) &= K[(x, Df(\theta, \theta)(x, y))] \\
&= (x, M^{-1}(Lx + My - Lx)) \\
&= (x, y).
\end{aligned}
$$

This establishes that K is the inverse of $Dh(\theta, \theta)$. Now **10.7** may be applied to conclude that there is an open set U in $X \times Y$ such that U contains (θ, θ), $V = h(U)$ is open in $X \times Y$ and again contains (θ, θ), and the restriction of h to U is a bijection of U onto V with continuous inverse in $C^1(V)$. Let this inverse be ϕ. We can express ϕ as

$$\phi(x, y) = (\phi_1(x, y), \phi_2(x, y)), \qquad (x, y) \in V,$$

where ϕ_1 and ϕ_2 map V into X and Y respectively. For any $(x, y) \in V$ we have

$$
\begin{aligned}
(x, y) = (h \circ \phi)(x, y) &= h(\phi_1(x, y), \phi_2(x, y)) \\
&= (\phi_1(x, y), f(\phi_1(x, y), \phi_2(x, y))).
\end{aligned}
$$

Thus for any $(x, y) \in V$, $\phi_1(x, y) = x$ and $f(x, \phi_2(x, y)) = y$. The function ϕ_2 now plays a crucial role in the manufacturing of g. Define the 'second coordinate projection' map $P : X \times Y \to Y$ by $P(x, y) = y$. By one of the exercises in chapter **10**, P is in $C^1(U)$. Also $\phi_2 = P \circ \phi$ and since ϕ is in $C^1(V)$, it follows from **10.3** that $\phi_2 \in C^1(V)$. Let

$$W = \{x \in X : (x, \theta) \in V\},$$

so that W is an open set in X which contains θ. Define $g : W \to Y$ by $g(x) = \phi_2(x, \theta)$ for $x \in W$. Then

$$g(\theta) = \phi_2(\theta, \theta) = (P \circ \phi)(\theta, \theta) = P(h^{-1}(\theta, \theta)) = P(\theta, \theta) = \theta.$$

Also, if $x \in W$ then $(x, \theta) \in V$ and so

$$f(x, g(x)) = f(x, \phi_2(x, \theta)) = \theta \quad \text{for all } x \text{ in } W.$$

Finally,

$$Dg(w)(x) = D\phi_2(w, \theta)(x, \theta)$$

for all $w \in W$ and $x \in X$. Hence $g \in C^1(W)$, and so g has all the required properties on the open set W.

(ii) The sets U and W as described in the proof of (i) suffice in this case. Firstly, $(\theta, \theta) \in U$. Secondly, if $(x, y) \in U$ and $f(x, y) = \theta$ then

$$h(x, y) = (x, f(x, y)) = (x, \theta).$$

Since $(x, y) \in U$, $h(x, y) \in V$ and so $x \in W$. It follows that

$$(x, y) = (\phi \circ h)(x, y) = \phi(x, \theta) = (x, \phi_2(x, \theta)) = (x, g(x)),$$

so that $y = g(x)$. On the other hand, if $(x, y) \in U$ and $y = g(x)$ then, since $x \in W$,

$$f(x, y) = f(x, g(x)) = \theta. \qquad \blacksquare$$

Exercises

1. The area of the unit disc $\{(s, t) : s^2 + t^2 \leq 1\}$ in $\mathbb{R}^2$ is known to be π. Calculate the areas of the elliptical discs
   (i) $\{(s, t) : s^2/4 + t^2/9 \leq 1\}$
   (i) $\{(s, t) : 2s^2 + 2st + 5t^2 \leq 1\}$.
   Note that $2s^2 + 2st + 5t^2 = (s + 2t)^2 + (s - t)^2$.

2. Let $B = \{(s, t) : s \geq 0, t \geq 0, 1 \leq s + t \leq 2\}$. Let $u = s + t$ and $v = t$ so that $(s, t) = \phi(u, v) = (u - v, v)$. Show that ϕ is bijective on $\mathbb{R}^2$, $J_\phi = 1$ and B is the image under ϕ of

$$C = \{(u, v) : 1 \leq u \leq 2, 0 \leq u \leq v\}.$$

Deduce that

$$\int_B (s + t) \, ds \, dt = \int_C u \, du \, dv = 7/3.$$

3. Evaluate

$$\int_1^3 \int_{s^2}^{s^2+1} st \, ds \, dt$$

directly, and by using the transformation $(s,t) \mapsto (u,v)$ where $(u,v) = (s, t - s^2)$.

4. Let the function $g : \mathbb{R}^2 \to \mathbb{R}$ be defined by

$$g(s,t) = s - t^2.$$

Show that there exist two continuous functions ϕ_1 and ϕ_2 such that

$$g(0, \phi_i(0)) = 0, \quad i = 1, 2.$$

5. Define $f : \mathbb{R}^2 \to \mathbb{R}$ by

$$f(s,t) = \begin{cases} s & t = 0 \\ s - t^3 \sin(1/t) & t \neq 0 \end{cases}$$

Show that there is an open set W containing $(0,0)$ such that $f \in C^1(W)$, but that it is impossible to find a function ϕ continuous on an open set V containing 0 in $\mathbb{R}$, such that $f(s, \phi(s)) = 0$ for $s \in V$.

Appendix
Countability

This short appendix deals very briefly with the topic of countability, which is one way of measuring the size of a set. At an elementary level, two sets A and B can be thought of as having the same size if there is a bijection $f : A \to B$. If this is the case, then A and B are said to be in *one-to-one* correspondence, each element a in A corresponding uniquely to the element $f(a)$ in B. The intuitive idea of a countable set is that it has the same size as the set $\mathbb{N}$ of natural numbers. However, such a definition would have the effect that finite sets are uncountable. This is not what we seek. We want the term countable to refer to sets which are somehow no bigger than $\mathbb{N}$.

Definition A.1 *A set A is countable if either A is the empty set, or if there is a mapping from $\mathbb{N}$ onto A.*

Lemma A.2 *If A is a finite set, then A is countable.*

Lemma A.3 *Every subset of a countable set is countable.*

Proof. Let A be a countable set, and f a mapping from $\mathbb{N}$ onto A. Take a non-empty set B contained in A, and fix $b \in B$. Define

$$g(n) = \begin{cases} f(n) & f(n) \in B \\ b & f(n) \notin B. \end{cases}$$

Then g maps $\mathbb{N}$ onto B, and so B is countable. ∎

Lemma A.4 *A countable set is either finite or in one-to-one correspondence with $\mathbb{N}$.*

Proof. Suppose A is infinite, and f maps $\mathbb{N}$ onto A. Define a mapping g recursively as follows. Set $g(1) = f(1)$. Let $g(2)$ be the first $f(r)$ which is different from $f(1)$. Such an $f(r)$ exists since A is infinite. Suppose $g(1), \ldots, g(n-1)$ have been defined where $n > 2$. Then $g(n)$ is taken to be the first number $f(r)$ that is different from all of the numbers $g(1), \ldots, g(n-1)$. The mapping g constructed in this way is a bijection from $\mathbb{N}$ to A. ∎

As was pointed out in chapter **4**, an infinite countable set A can be written as

$$A = \{a_1, a_2, a_3 \ldots .\},$$

where a_n is the image of $n \in \mathbb{N}$ under the bijection whose existence was demonstrated in lemma **A.4**. Another way to describe a countable set in the light of lemma **A.4** is as follows. If the set A is countable, then the elements can be labelled as the first element, the second element and so on.

Lemma A.5 *The union of a countable family of countable sets is itself a countable set.*

Proof. If a family of sets is countable, then it may be written $A_1, A_2, A_3, \ldots$. Since each A_i is countable, we may write

$$A_i = \{a_{i1}, a_{i2}, \ldots\}, \quad i = 1, 2, \ldots.$$

Now define the function $f : \mathbb{N} \to \cup_{i=1}^{n} A_i$ by $f(1) = a_{11}$, $f(2) = a_{21}$, $f(3) = a_{12}$, and so on. Then f is a bijection from $\mathbb{N}$ to $\cup_{i=1}^{n} A_i$, and so this set is countable. ∎

Corollary A.6 *If A and B are countable sets, then so is $A \times B$.*

Proof. Since B is countable, it may be written $\{b_1, b_2, \ldots\}$. Then

$$A \times B = \cup_{n=1}^{\infty} A \times \{b_n\}.$$

Clearly each set $A \times \{b_n\}$ is countable, and so $A \times B$ is countable by lemma **A.5**. ∎

Corollary A.7 *The set Q consisting of all rational numbers is countable.*

Proof. The set $\mathbb{Z}$ is countable, and so by **A.6**, $\mathbb{Z} \times \mathbb{N}$ is countable. Hence there exists a mapping g from $\mathbb{N}$ onto $\mathbb{Z} \times \mathbb{N}$. Now the mapping $f : \mathbb{Z} \times \mathbb{N} \to \mathbb{Q}$ given by $f(p, q) = p/q$ is a surjective mapping. Thus $g \circ f$ is a mapping from $\mathbb{N}$ onto $\mathbb{Q}$, and so $\mathbb{Q}$ is countable. ∎

Our treatment of countability would be incomplete if we did not give at least one example of an uncountable set.

Lemma A.8 *The set of all real numbers is uncountable.*

Proof. In fact the set $[0, 1]$ is uncountable, as we shall now show. Let $\{x_1, x_2, \ldots\}$ be any countable set in $[0, 1]$. We shall show that this set cannot be the whole of $[0, 1]$. Write x_n as a decimal expansion,

$$x_n = 0 \cdot a_{n1}a_{n2}\ldots, \quad n = 1, 2, \ldots.$$

Define

$$y_n = \begin{cases} 1 & a_{nn} \neq 1 \\ 2 & a_{nn} = 1. \end{cases}$$

and set

$$y = 0.y_1 y_2 \ldots = \sum_{n=1}^{\infty} \frac{y_n}{10^n}.$$

For each n, $y \neq x_n$ since $y_n \neq a_{nn}$, and so $\{x_1, x_2, \ldots\}$ is not the whole of $[0, 1]$. Thus $[0, 1]$ is uncountable. ∎

We conclude with a simple result which was needed in chapter **5**.

Lemma A.9 *Let A and B be countable subsets of a linear space X. Then*

$$A + B = \{a + b : a \in A, \, b \in B\}$$

is countable.

Proof. Since A and B are countable, $A \times B$ is countable by **A.6**. Thus there is a mapping f from $\mathbb{N}$ onto $A \times B$. Define $g : A \times B \to A + B$ by $g(a, b) = a + b$. Then g maps $A \times B$ onto $A + B$, and so $g \circ f$ is a mapping from $\mathbb{N}$ onto $A + B$. Hence $A + B$ is countable. ∎

References and suggested reading

The text of this book contains a small number of references. They are listed first.

[1] Banach S., *Théorie des Opérations Linéaires*, Monografje Matematyczne, Warsaw, 1932. Reprinted by Chelsea, New York, 1955.

[2] Bielecki, A., *Une remarque sur la méthode de Banach-Cacciopoli-Tikhonov*, Bull. Acad. Plon. Sci. IV (1956), 208-215.

[3] Burkhill, J. C., *The Theory of Ordinary Differential Equations*, Oliver and Boyd, 1962.

[4] Rosenlicht, M., *Introduction to Analysis*, Scott and Foresman, 1968.

[5] Spivak, M., *Calculus*, W.A. Benjamin, Inc. London, 1967.

As mentioned in the introduction, it is my hope that the reader will now feel able and eager to tackle other topics in analysis. The following short list should be helpful.

Simmons, G. F., *Introduction to Topology and Modern Analysis*, McGraw-Hill, New York, 1963.

Jameson, G. J. O. J., *Topology and Normed Spaces*, Chapman and Hall, London, 1974.

Brown, A. L. and Page, A. *Elements of Functional Analysis*, Van Nostrand Reinhold, London, 1970.

Rudin, W., *Real and Complex Analysis*, McGraw-Hill, New York, 1966.

Taylor, A. E., *Introduction to Functional Analysis*, Wiley, New York, 1958.

Diestel, J., *Sequences and Series in Banach Spaces*, Springer-Verlag, Heidelberg, 1988.

These books all take the subject of abstract analysis quite a bit further. Simmons is a nice elementary text with plenty of motivation and discussion, treating topology as well as linear analysis. The book by Jameson dwells at some length on the classical theory of Banach spaces. It is very readable. Rudin gives an introduction to the theory of topological spaces and measure spaces side by side. I don't know a nicer introduction to the two theories. The book by Diestel shouldn't really be in the list at all. It is quite hard in places, covering some rather recent results in Banach space theory. Nevertheless, it is an inspiring book, and some of the chapters are not too hard. The determined reader should be able to make progress with this book, and it gives an exciting overview of some of the more modern developments. Finally, the reader who is interested in the historical aspects of the subject will find the book by Banach[1] readily accessible.

The reader may, on the other hand, want to look deeper into an area of classical analysis, particularly analysis in $\mathbb{R}^n$. There are many useful and interesting results which would have taken us too far afield in this book. Two good references are

Bartle, R., *An Introduction to Real Analysis*, 2nd ed., John Wiley and Sons, Inc., New York, 1976

Apostol, T. M., *Mathematical Analysis*, Addison-Wesley, Reading, Mass., 1957.

Index

T - #0179 - 071024 - C0 - 229/152/11 - PB - 9780412310904 - Gloss Lamination